Preface

This workbook is a supplement to the textbook *Engineering Mechanics: Dynamics*. As a result, the problems in this book are arranged in the same order as those presented in the textbook. Here the solution to the problems is only partially complete. The key equations, which stress the important fundamentals of the problem solution, must be supplied in the space provided. There is no need for calculations, however, since all the answers are given in the back of the book.

It is suggested that these problems by solved just after the theory and example problems covering the corresponding topic have been studied in the textbook. If an honest effort is made at completing and understanding the solution to these problems, it will serve to build confidence in applying the theory to the textbook problems. Furthermore, these problems provide an excellent review of the subject matter, which can then be used when preparing for exams.

I would greatly appreciate hearing from you if you have any comments or suggestions regarding the contents of this work.

Russell Charles Hibbeler
hibbeler@bellsouth.net

PRACTICE PROBLEMS WORKBOOK

ENGINEERING MECHANICS

DYNAMICS

TENTH EDITION

R. C. Hibbeler

Prentice Hall

Upper Saddle River, New Jersey 07458

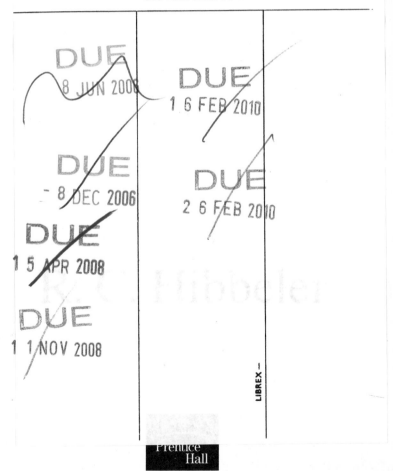

Executive Editor: *Eric Svendsen*
Associate Editor: *Dee Bernhard*
Supplement Cover Manager: *Daniel Sandin*
Executive Managing Editor: *Vince O'Brien*
Managing Editor: *David A. George*
Production Editor: *Barbara A. Till*
Buyer: *Ilene Kahn*

PEARSON
Prentice Hall

Published by Pearson Prentice Hall
Pearson Education, Inc.
Upper Saddle River, NJ 07458

Printed in the United States of America
10 9 8 7 6 5 4 3 2 1

ISBN 0-13-141679-0

Pearson Education Ltd., *London*
Pearson Education Australia Pty. Ltd., *Sydney*
Pearson Education Singapore, Pte. Ltd.
Pearson Education North Asia Ltd., *Hong Kong*
Pearson Education Canada, Inc., *Toronto*
Pearson Educación de Mexico, S.A. de C.V.
Pearson Education—Japan, *Tokyo*
Pearson Education Malaysia, Pte. Ltd.
Pearson Education, Inc., *Upper Saddle River, New Jersey*

Contents

12 Kin[illegible]cle

Rectilinear Kinematics

12 - 1. A car is traveling at a speed of 8 m/s when the brakes are suddenly applied, causing a constant deceleration 1 m/s^2. Determine the time required to stop the car and the distance traveled before stopping.

Solution

Since the deceleration is constant, determine the time *t* using

$$(\overset{+}{\rightarrow})\ v = v_0 + a_c t$$

$t = 8$ s **Ans.**

Determine the distance traveled using

$$(\overset{+}{\rightarrow})\ v^2 = v_0^2 + 2a_c(s - s_0)$$

$s = 32$ m **Ans.**

Also, using the time *t* computed above the following equation can be used to determine the distance.

$$(\overset{+}{\rightarrow})\ s = s_0 + v_0 t + \frac{1}{2} a_c t^2$$

$s = 32$ m **Ans.**

12 - 2. A particle is moving along a straight line through a fluid medium such that its speed is $v = (2t)$ m/s, where t is in seconds. If it is released from rest at $s = 0$, determine its position and acceleration when $t = 3$ s.

Solution

Since v is a function of t, then s can be determined from

$$v = \frac{ds}{dt} = 2t$$

Set up the definite integral to determine $s = f(t)$.

____

$$s = t^2$$

At $t = 3$ s,

$$s = (3)^2 = 9 \text{ m}$$ **Ans.**

Also, since v is a function of t, then determine a using

$$a = \frac{dv}{dt}$$

$a =$ __________ **Ans.**

12 - 3. A ball is thrown vertically upward from the top of a building with an initial velocity of $v_A = 12$ m/s. Determine (a) how high above the top of the building the ball will travel before it stops at B, (b) the time t_{AB} it takes to reach its maximum height, and (c) the total time t_{AC} needed for it to reach the ground at C from the instant it is released.

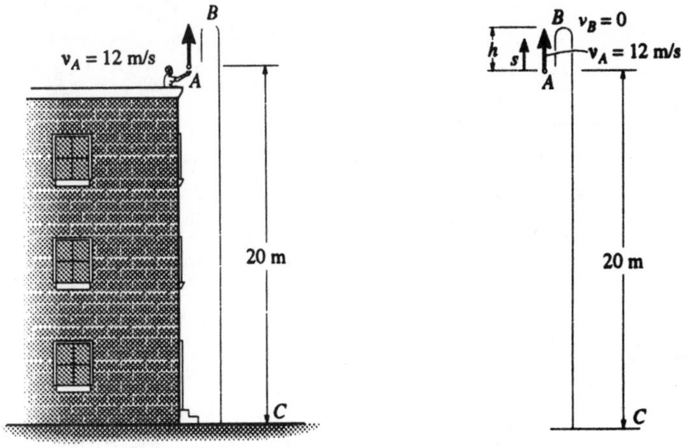

Solution

a) Determine the height of h.

$$(+\uparrow)v_B^2 = v_A^2 + 2a_c(s_B - s_A)$$

$h = 7.34$ m **Ans.**

b) Determine the time t_{AB}.

$$(+\uparrow)v_B = v_A + a_c t$$

$t_{AB} = 1.22$ s **Ans.**

c) Determine the time t_{AC}.

$$(+\uparrow)s_C = s_A + v_A t + \frac{1}{2}a_c t^2$$

Solving for the positive root,

$t_{AC} = 3.58$ s **Ans.**

12 - 4. A small metal particle passes downward through a fluid medium while being subjected to the attraction of a magnetic field such that its position is observed to be $s = (15t^3 - 3t)$ mm, where t is in seconds. Determine (a) the particle's displacement from $t = 2$ s to $t = 4$ s, and (b) the velocity and acceleration of the particle when $t = 5$ s.

Solution

a) When

$t = 2$ s, $s =$ ____ $= 114$ mm

$t = 4$ s, $s =$ ____ $= 948$ mm

The displacement is therefore

$\Delta s =$ ______________________ $= 834$ mm **Ans.**

b) Using s as a function of time we can find v from

$$v = \frac{ds}{dt} = \text{\_\_\_\_\_\_\_\_\_\_\_\_\_\_\_\_\_\_\_\_}$$

When

$t = 5$ s, $v = 1122$ mm/s **Ans.**

Using v as a function of time we can find a from

$$a = \frac{dv}{dt} = \text{\_\_\_\_\_\_\_\_\_\_\_\_\_\_\_\_\_\_\_\_}$$

When

$t = 5$ s, $a = 450$ mm/s^2 **Ans.**

12 - 5. A car, initially at rest, moves along a straight road with constant acceleration such that it attains a velocity of 20 m/s when $s_1 = 50$ m. Then after being subjected to *another* constant acceleration, it attains a final velocity of 30 m/s when $s_2 = 200$m. Determine the average velocity and average acceleration of the car for the entire 200 m displacement.

Solution

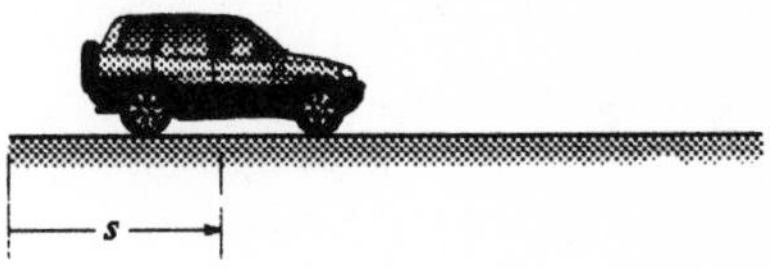

Determine the first acceleration using

$$(\overset{+}{\rightarrow})\, v_1^2 = v_0^2 + 2a_1(s_1 - s_0)$$

____

$$a_1 = 4 \text{ m/s}^2$$

Determine the second acceleration using

$$(\overset{+}{\rightarrow})\, v_2^2 = v_1^2 + 2a_2(s_2 - s_1)$$

____

$$a_2 = 1.25 \text{ m/s}^2$$

Determine the first time period using

$$(\overset{+}{\rightarrow})\, v_1 = v_0 + a_1 t_1$$

____

$$t_1 = 5 \text{ s}$$

Determine the second time period using

$$(\overset{+}{\rightarrow})\, v_2 = v_1 + a_2 t_2$$

____

$$t_2 = 8 \text{ s}$$

Thus,

$$v_{avg} = \frac{\Delta s}{\Delta t} = \_\_\_\_\_\_\_\_\_\_ = 11.8 \text{ m/s}$$ **Ans.**

$$a_{avg} = \frac{\Delta v}{\Delta t} = \_\_\_\_\_\_\_\_\_\_ = 1.76 \text{ m/s}^2$$ **Ans.**

Erratic Motion

12 - 6. A car travels up a hill with the speed shown. Determine the total distance the car moves until it stops (t = 60 s). Plot the $a-t$ graph.

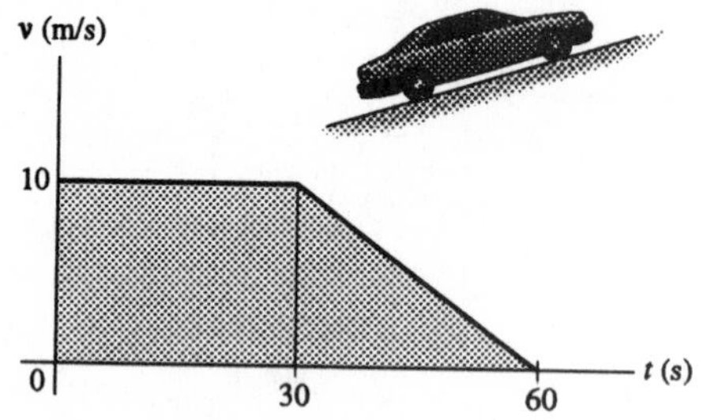

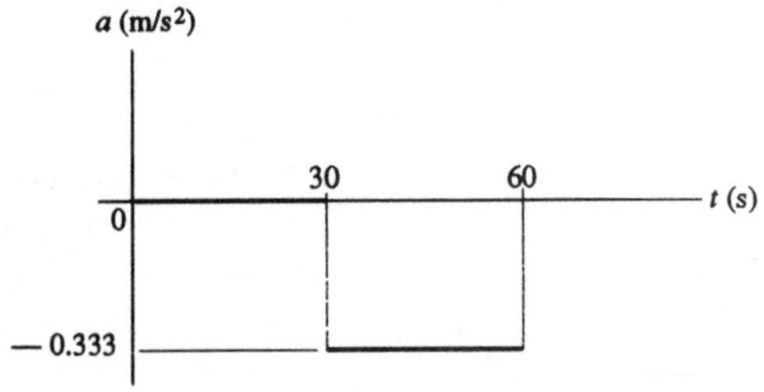

Solution

Since $\Delta s = \int v dt$, then the total distance traveled is the area under the v - t graph $0 \leq t \leq 60$ s.

$A_1 + A_2 =$ ______________________

$s = 450$ m **Ans.**

The slope of the $v-t$ graph is $a = dv/dt = 0$ for $0 \leq t < 30$ s, and $a = dv/dt = -0.333$ for $30 \text{ s} < t \leq 60$ s. Therefore the $a-t$ graph is shown below.

12 - 7. A race car starting from rest moves along a straight track with an acceleration as shown. Determine the time t for the car to reach a speed of 50 m/s and construct the $v-t$ graph that describes the motion.

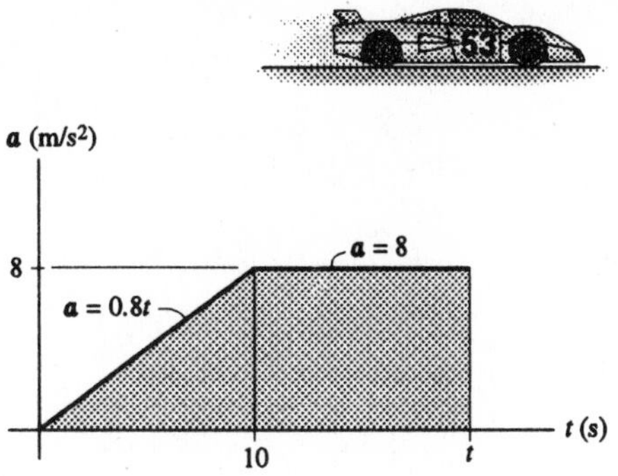

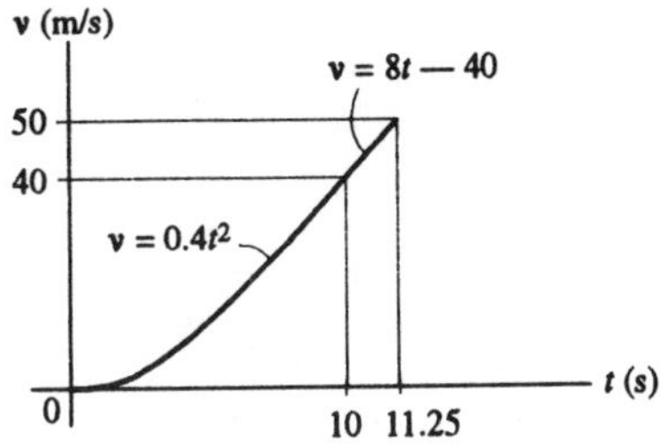

Solution

From the graph, for $0 \le t < 10$ s set up the definite integral to determine $v = f(t)$.

____

$$v = 0.4t^2$$

When $t = 10$s, $v =$ ______________________________ $= 40$ m/s

For $t > 10$ s, set up the integrals to determine $v = f(t)$.

____

$$v - 40 = 8t - 80$$
$$v = 8t - 40$$

When $v = 50$ m/s

$t = 11.25$ **Ans.**

Also, since $\Delta v = \int a dt$, this problem can be solved by finding the area under the $a-t$ graph. We require

$$50 - 0 = \frac{1}{2}(10)(8) + 8(t - 10)$$
$$t = 11.25 \text{ s}$$

12 - 8. A car starts from rest and travels along a straight road having the acceleration shown. Plot the $v-t$ and $s-t$ graphs which describe the motion.

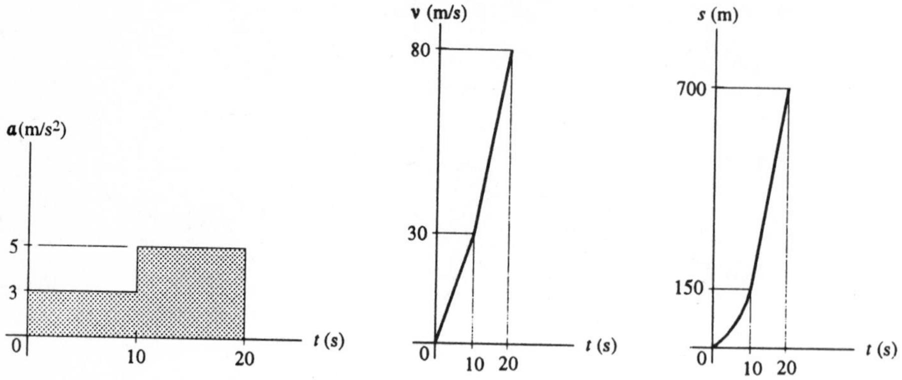

Solution

Since $\Delta v = \int a dt$, the constant lines of the $a-t$ graph become sloping lines for the $v-t$ graph. Since $v = 0$ when $t = 0$, the numerical values at any time are calculated as the area under the $a-t$ graph.

When $t = 10$ s, $v =$ ______________________ $= 30$ m/s

When $t = 20$ s, $v =$ ______________________ $= 80$ m/s

Since $\Delta s = \int v dt$, the sloping lines of the $v-t$ graph become parabolic curves for the $s-t$ graph. Since $s = 0$ when $t = 0$, the numerical values at any time are calculated from the total area under the $v-t$ graph.

When $t = 10$ s, $s =$ ______________________

$s = 150$ m

When $t = 20$ s, $s =$ ______________________

$s = 700$ m

12 - 9. From experimental data, the motion of the motorcycle is defined by the $v-t$ graph shown. Construct the $s-t$ and $a-t$ graphs for the motion.

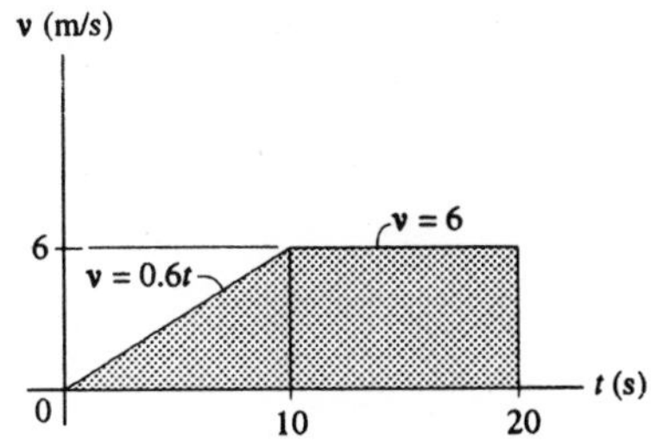

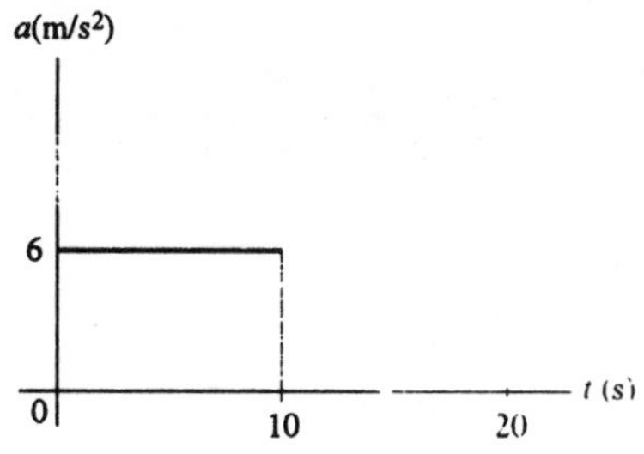

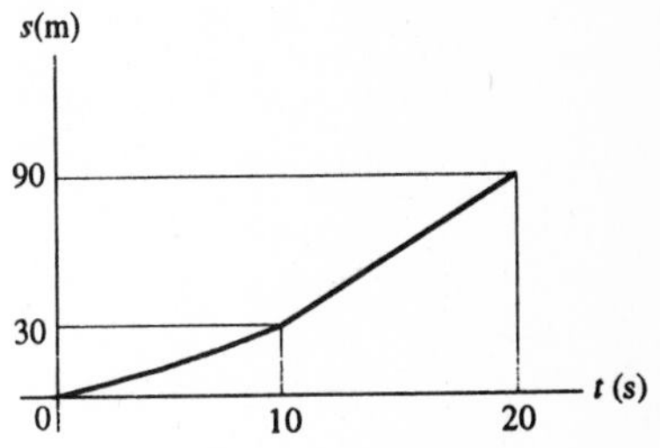

Solution

Since $a = dv/dt$, values of a are determined from the slope of the $v-t$ graph.

$$0 \le t < 10\ \text{s};\ a = \frac{\Delta v}{\Delta t} = \underline{\hspace{8cm}} = 0.6\ \text{m/s}^2$$

$$10\ \text{s} < t \le 20\ \text{s};\ a = \frac{\Delta v}{\Delta t} = \underline{\hspace{8cm}} = 0$$

Since $\Delta s = \int v dt$, motorcycle's position when $t_1 = 10$ s and $t_2 = 20$ s is determined from the area under the $v-t$ graph.

$$s_1 = \underline{\hspace{8cm}} = 30\ \text{m}$$

$$s_2 = \underline{\hspace{8cm}} = 90\ \text{m}$$

Set up the definite integrals to determine the equations of each segment of the $s-t$ graph.

$$0 \le t < 10\ \text{s};\ ds = vdt;\ \underline{\hspace{8cm}}$$

$$s = 0.3t^2$$

$$10\ \text{s} < t \le 20\ \text{s};\ ds = vdt;\ \underline{\hspace{8cm}}$$

$$s = 6t - 30$$

12 - 10. The $v-s$ graph for a truck is shown. Determine the acceleration of the truck at s = 100 m and s = 175 m.

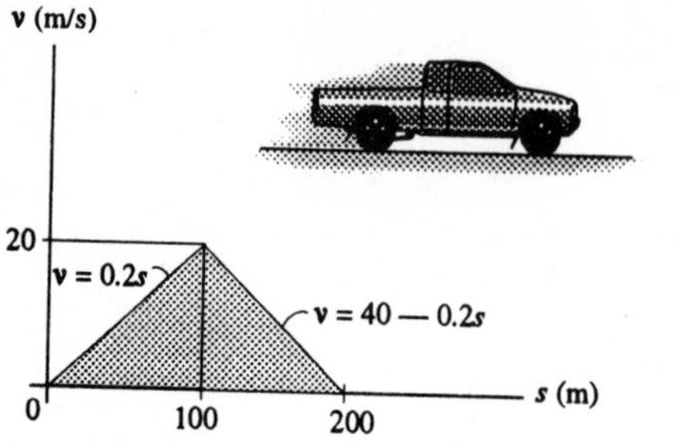

Solution

For $0 \leq s < 100$ m, set up the equation to determine $a = f(s)$ using $vdv = ads$.

____

$$a = 0.04s$$

At s = 100 m,

$$a = 0.04(100) = 4 \text{ m/s}^2$$ **Ans.**

For $100 \text{ m} < s \leq 200$ m, set up the equation to determine $a = f(s)$.

____

$$a = 0.04s - 8$$

At s = 175 m,

$$a = 0.04(175) - 8$$

$$a = -1 \text{ m/s}^2$$ **Ans.**

Curvilinear Motion: Rectangular Components

12-11. The particle travels along the path defined by the parametric equations $x = (0.125t^2)$ m and $y = (0.03t^3)$ m, where t is in seconds. When $t = 4$ s, determine the particle's distance from the origin, its speed, and the magnitude of its acceleration.

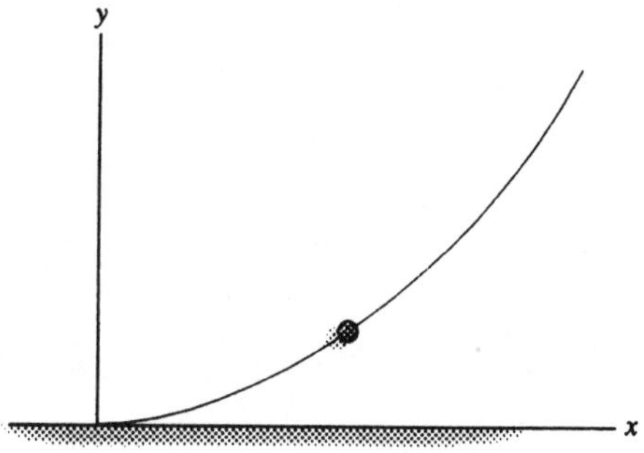

Solution

When $t = 4$ s,

$x =$ ____________________ $= 2$ m

$y =$ ____________________ $= 1.92$ m

$d =$ ____________________ $= 2.77$ m **Ans.**

$v_x = \dot{x} =$ ____________________ $= 1$ m/s

$v_y = \dot{y} =$ ____________________ $= 1.44$ m/s

$v = \sqrt{(1)^2 + (1.44)^2} = 1.75$ m/s **Ans.**

$a_x = \ddot{x} =$ ____________________ $= 0.250$ m/s^2

$a_y = \ddot{y} =$ ____________________ $= 0.720$ m/s^2

$a = \sqrt{(0.250)^2 + (0.720)^2} = 0.762$ m/s^2 **Ans.**

12 - 12. For a short time the particle moves along the parabolic path $y = (18 - 2x^2)$m. If motion along the ground is $x = (4t - 3)$ m, where t is in seconds, determine the magnitudes of the particle's velocity and acceleration when $t = 1$s.

Solution

Take the time derivatives as follows :

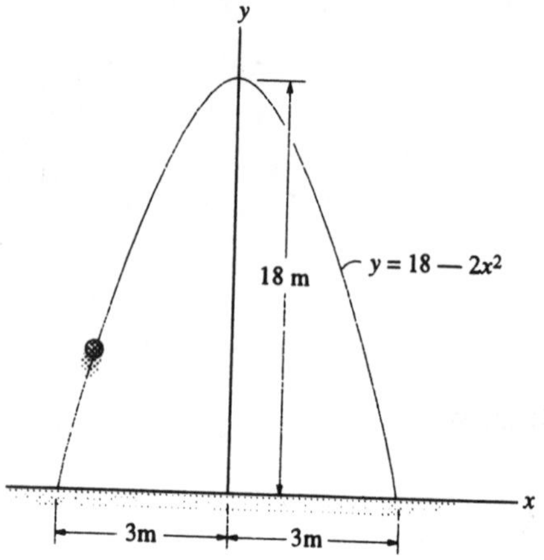

$x = 4t - 3$

$v_x = \dot{x} =$ ________________________

$a_x = \ddot{x} =$ ________________________

$y = 18 - 2x^2$

$v_y = \dot{y} =$ ________________________

$a_y = \ddot{y} =$ ________________________

When $t = 1$ s,

$v_x = 4\text{m/s}$

$v_y = -4(1)(4) = -16\text{ m/s}$

$v = \sqrt{(4)^2 + (-16)^2} = 16.5\text{ m/s}$ **Ans.**

$a_x = 0$

$a_y = [-4(4)^2 - 4(1)(0)] = -64\text{ m/s}^2$

$a = \sqrt{(0)^2 + (-64)^2} = 64.0\text{ m/s}^2$ **Ans.**

12 - 13. The motorcyclist attempts to jump over a series of cars and trucks and land smoothly on the other ramp, i.e., such that his velocity is tangent to the ramp at B. Determine the launch speed v_A necessary to make the jump.

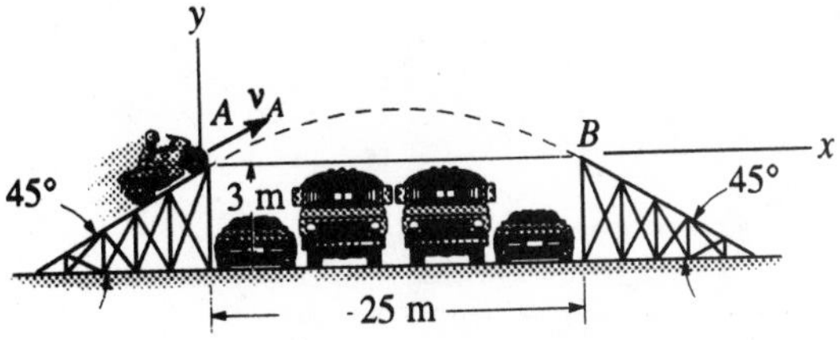

Solution

At A,

$(v_A)_x =$ ______________________

$(v_A)_y =$ ______________________

Apply the following kinematic equations in the horizontal and vertical directions.

$(\xrightarrow{+})(s_B)_x = (s_A)_x + (v_A)_x t$

____

$(+\uparrow)(s_B)_y = (s_A)_y + (v_A)_y t + \frac{1}{2} a_c t^2$

____

Solving,

$v_A = 16.8\ \text{m/s}$ **Ans.**

$t = 1.72\ \text{s}$

12 - 14. It is observed that the ball when thrown at A reaches its maximum height h at B in $t = 2\text{s}$. Determine the speed v_A at which it was thrown, the angle of release θ, and the height h.

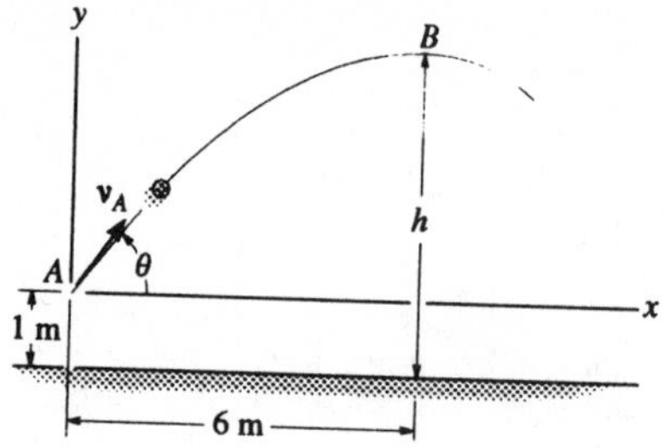

Solution

The origin of coordinates is established at A. Apply the following equations :

$$(\overset{+}{\rightarrow})(s_B)_x = (s_A)_x + (v_A)_x t \qquad (1)$$

____

$$(+\uparrow)(v_B)_y^2 = (v_A)_y^2 + 2a_c[(s_B)_y - (s_A)_y] \qquad (2)$$

____

$$(+\uparrow)(v_B)_y = (v_A)_y + a_c t \qquad (3)$$

____

To solve, rearrange Eq. 3 and divide it by Eq. 1, to obtain $\tan\theta = 6.54$.

Then,

$\theta = 81.3°$ **Ans.**

$v_A = 19.8 \text{ m/s}$ **Ans.**

$h = 20.6 \text{ m}$ **Ans.**

12-15. A ball is launched perpendicular to the incline with a velocity of $v_A = 15$ m/s. Determine the distance R where it strikes the plane at B.

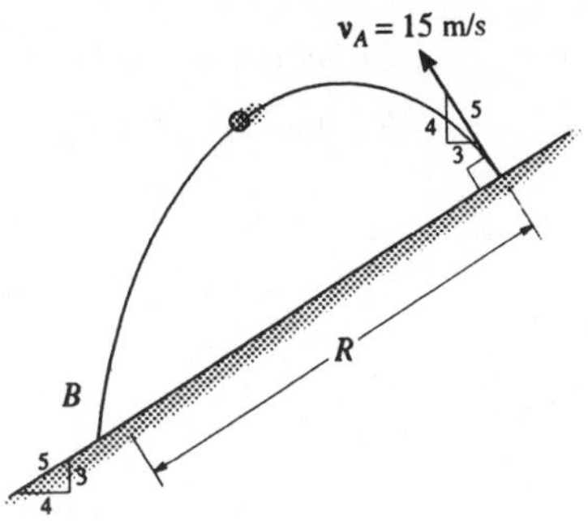

Solution

The origin of the coordinates is established at A. Then

$(v_A)_x =$ ________________________________ $= 9$ m/s $\leftarrow$

$(v_A)_y =$ ________________________________ $= 12$ m/s $\uparrow$

Write two kinematic equations in terms of R and t.

$(\overset{+}{\leftarrow})$ ____

$(+\uparrow)$ ____

Eliminating R,

$$-6.75t = 12t - 4.905t^2$$

Solving

$$t = 3.82 \text{ s}$$

Thus,

$$R = 43.0 \text{ m}$$ **Ans.**

Curvilinear Motion:
Normal and Tangential Components

12 - 16. A car is traveling along a circular path having a radius of 60 m. Determine the magnitude of the car's acceleration if at a given instant its speed is $v = 7$ m/s and the rate of decrease in speed is $\dot{v} = 3$ m/s^2.

Solution

$a_n =$ ______________________ $= 0.817$ m/s^2

$a_t =$ ______________________

$a =$ ______________________ $= 3.11$ m/s^2 **Ans.**

12 - 17. A train travels along a horizontal circular curve that has a radius of 200 m. If the speed of the train is uniformly increased from 8 m/s to 10 m/s in 5 s, determine the magnitude of the acceleration at the instant the speed of the train is 9 m/s.

Solution

$a_t =$ ______________________ $= 0.4$ m/s^2

$a_n =$ ______________________ $= 0.405$ m/s^2

$a =$ ______________________ $= 0.569$ m/s^2 **Ans.**

12 - 18. A particle is traveling down along the parabola $y = \frac{1}{4}x^2$. When it reaches point A it has a speed of 8m/s which is increasing at $\dot{v} = 4\ \text{m/s}^2$ Determ ine the magnitude of the acceleration of the particle at this instant.

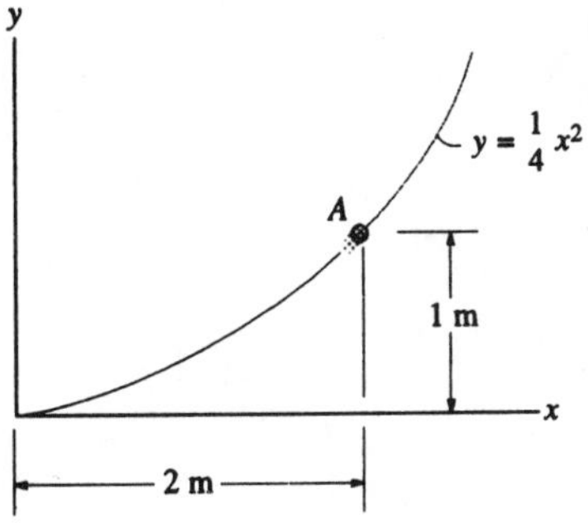

Solution

$$y = \frac{1}{4}x^2$$

$$\left.\frac{dy}{dx}\right|_{x=2} = \_\_\_\_\_\_\_\_\_\_\_\_\_\_\_\_\_\_\_\_$$

$$\frac{d^2y}{dx^2} = \_\_\_\_\_\_\_\_\_\_\_\_\_\_\_\_\_\_\_\_$$

$$\rho = \left|\frac{[1+(\frac{dy}{dx})^2]^{3/2}}{\frac{d^2y}{dx^2}}\right| = \left|\frac{[1+(1)^2]^{3/2}}{\frac{1}{2}}\right| = 5.66\ \text{m}$$

$$a_t = \_\_\_\_\_\_\_\_\_\_\_\_\_\_\_\_\_\_\_\_$$

$$a_n = \_\_\_\_\_\_\_\_\_\_\_\_\_\_\_\_\_\_\_\_ = 11.31\ \text{m/s}^2$$

$$a = \sqrt{(4)^2 + (11.31)^2} = 12.0\ \text{m/s}^2$$ **Ans.**

12 - 19. When the car is at A its speed increases along the verticle circular path at the rate of $\dot{v} = (0.2t)\ \text{m/s}^2$, wher t is in seconds. If it starts from rest at A, determine its velocity and acceleration when it reaches B.

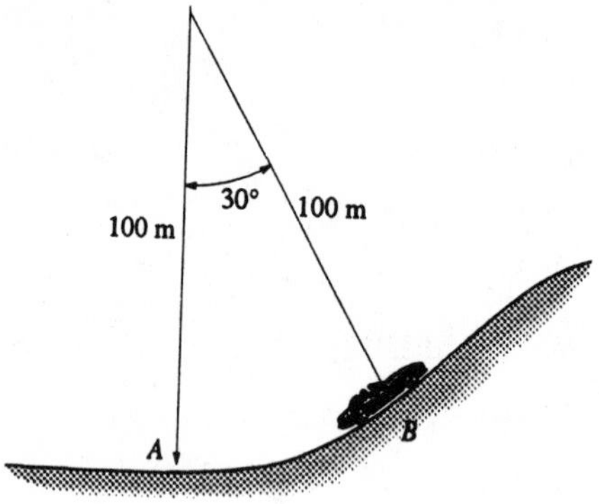

Solution

Since $v = 0$ when $t = 0$, v as a function of time is

____

Since $s = 0$ when $t = 0$, s as a function of time is

____

The length of path is

$s_{AB} =$ ______________________________ $= 52.36\ \text{m}$

The time to reach B is

$t_B =$ ______________________________ $= 11.62\ \text{s}$

Thus

$v_B =$ ______________________________ $= 13.5\ \text{m/s}$ **Ans.**

$(a_t)_B =$ ______________________________ $= 2.325\ \text{m/s}^2$

$(a_t)_B =$ ______________________________ $= 1.826\ \text{m/s}^2$

$a_B =$ ______________________________ $= 2.96\ \text{m/s}^2$ **Ans.**

12 - 20. A package is dropped from the plane which is flying with a constant horizontal velocity of $v_A = 50$ m/s. Determine the tangential and normal components of acceleration and the radius of curvature of the path of motion (a) at the moment the package is released at A, where it has a horizontal velocity of $v_A = 50$ m/s, and (b) just before it strikes the ground at B.

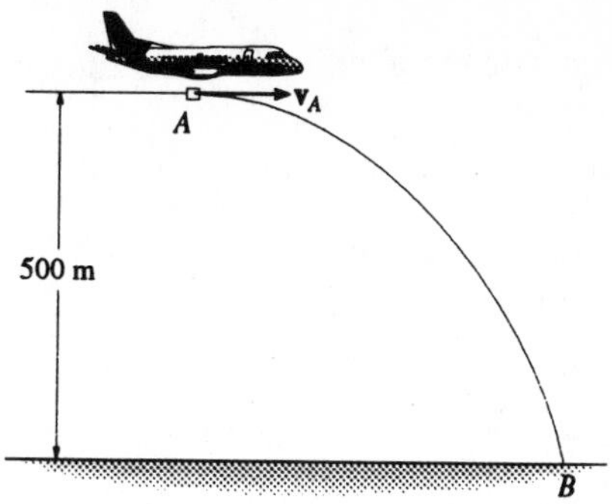

Solution

Since the acceleration is 9.81 m/s², then

a) $(a_n)_A =$ ____________________ **Ans.**

$(a_t)_A =$ ____________________ **Ans.**

Knowing v_A and $(a_n)_A$, then

$\rho_A =$ ____________________ = 255 m **Ans.**

b) $(v_B)_x =$ ____________________

To determine $(v_B)_y$, apply

$$(+\downarrow)(v_B)_y^2 = (v_A)_y^2 + 2a_c[(s_B)_y - (s_A)_y]$$

____

$(v_B)_y = 99.05$ m/s

$v_B =$ ____________________ = 110.95 m/s

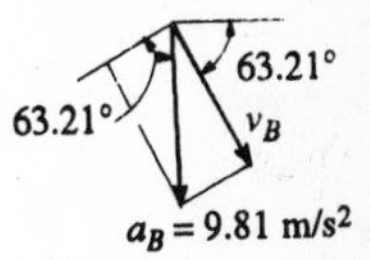

$$\theta = \tan^{-1}\frac{99.05}{50} = 63.21°$$

$(a_n)_B =$ ____________________ = 4.42 m/s² **Ans.**

$(a_t)_B =$ ____________________ = 8.76 m/s² **Ans.**

$\rho_B =$ ____________________ = 2.79 km **Ans.**

Curvilinear Motion: Cylindrical Components

12-21. If $\theta = f(t)$, use the chain rule of calculus and take the time derivatives of the following expressions to determine $\dot{r}$ and $\ddot{r}$.

$r = e^{\theta}$

$\dot{r} =$ ____________________

$\ddot{r} =$ ____________________

$r = 2\theta^2 + 8\sin\theta$

$\dot{r} =$ ____________________

$\ddot{r} =$ ____________________

$r^2 = 4\sin 2\theta$

$2r\dot{r} =$ ____________________

$2\dot{r}^2 + 2r\ddot{r} =$ ____________________

12 - 22. A car is traveling along the circular curve of radius $r = 100$ m. At the instant shown, its angular rate of rotation is $\dot{\theta} = 0.4$ rad/s, which is increasing at the rate of $\ddot{\theta} = 0.2$ rad/s^2. Determine the magnitudes of the car's velocity and acceleration at this instant.

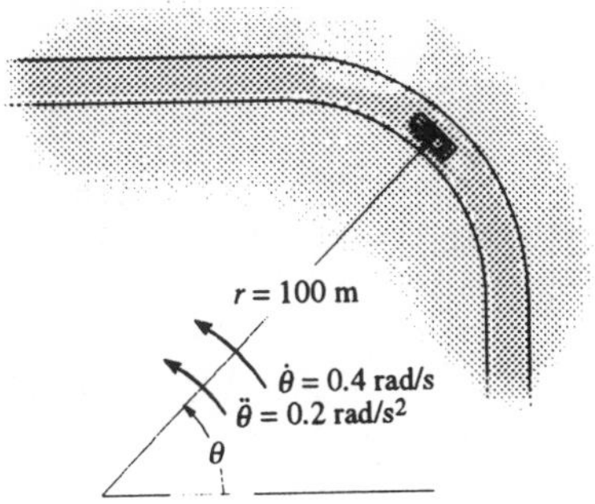

Solution

$\dot{\theta} = 0.4$ rad/s

$\ddot{\theta} =$ ______________________

$r = 100$ m

$\dot{r} =$ ______________________

$\ddot{r} =$ ______________________

$v_r = \dot{r} = 0$

$v_\theta = r\dot{\theta} = 100(0.4) = 40$ m/s

$v =$ ______________________ **Ans.**

$a_r = \ddot{r} - r\dot{\theta}^2 = 0 - 100(0.4)^2 = -16$ m/s^2

$a_\theta = r\ddot{\theta} + 2\dot{r}\dot{\theta} = 100(0.2) + 2(0)(0.4) = 20$ m/s^2

$a =$ ______________________ $= 25.6$ m/s^2 **Ans.**

12 - 23. For a short time the positon of a roller - coaster car along its path is defined by the equations $r = 25$ m, $\theta = (0.3t)$ rad, and $z = (-8\cos\theta)$ m, where t is in seconds. Determine the components of the car's velocity and acceleration when $t = 4$ s.

Solution

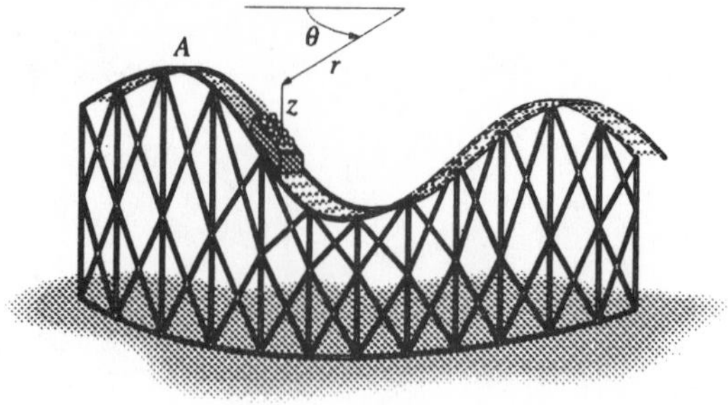

$r =$ ________________________________

$\dot{r} =$ ________________________________

$\ddot{r} =$ ________________________________

$\theta =$ ________________________________

$\dot{\theta} =$ ________________________________

$\ddot{\theta} =$ ________________________________

$z =$ ________________________________

$\dot{z} =$ ________________________________

$\ddot{z} =$ ________________________________

$v_r = \dot{r} = 0$ **Ans.**

$v_\theta = r\dot{\theta} = 25(0.3) = 7.5\text{ m/s}$ **Ans.**

$v_z = \dot{z} = 8\sin[0.3(4)](0.3) = 2.24\text{ m/s}$ **Ans.**

$a_r = \ddot{r} - r\dot{\theta}^2 = 0 - 25(0.3)^2 = 2.25\text{ m/s}^2$ **Ans.**

$a_\theta = r\ddot{\theta} + \dot{r}\dot{\theta} = 0 + 0 = 0$ **Ans.**

$a_z = \ddot{z} = 0 + 8\cos[0.3(4)](0.3)^2 = 0.261\text{ m/s}^2$ **Ans.**

12 - 24. The slotted link is pinned at O, and as a result of rotation it drives the peg P along the horizontal guide. Determine the magnitudes of the velocity and acceleration of P as a function of θ if $\theta = (3t)$ rad, where t is in seconds.

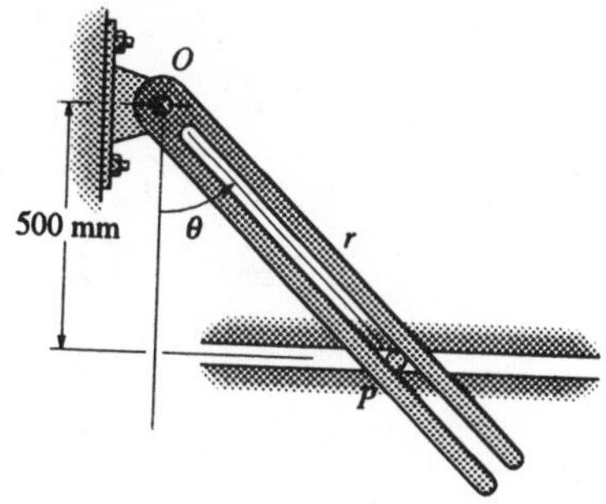

Solution

Here $r = (500/\cos\theta)\text{mm} = (500\sec\theta)\text{mm}$.

In the following use the identity $1 + \tan^2\theta = \sec^2\theta$ and recall $d(\sec\theta) = \sec\theta\tan\theta\, d\theta$, and $d(\tan\theta) = \sec^2\theta\, d\theta$.

$\dot{\theta} =$ ______________________________

$\ddot{\theta} =$ ______________________________

$\dot{r} =$ ______________________________ $= 1500\sec\theta\tan\theta$

$\ddot{r} =$ ______________________________ $= 4500(\sec\theta)(2\tan^2\theta + 1)$

$$v_r = \dot{r} = 1500\sec\theta\tan\theta$$

$$v_\theta = r\dot{\theta} = 1500\sec\theta$$

$$v = \sqrt{(1500)^2\sec^2\theta\tan^2\theta + (1500)^2\sec^2\theta}$$

$$v = 1500\sec^2\theta \text{ mm/s}$$ **Ans.**

$$a_r = \ddot{r} - r\dot{\theta}^2 = 4500\sec\theta(2\tan^2\theta + 1) - 500\sec\theta(3)^2 = 9000\sec\theta\tan^2\theta$$

$$a_\theta = r\ddot{\theta} + 2\dot{r}\dot{\theta} = 0 + 2(1500\sec\theta\tan\theta)(3) = 9000\sec\theta\tan\theta$$

$$a = 9000\sqrt{\sec^2\theta\tan^4\theta + \sec^2\theta\tan^2\theta}$$

$$a = 9000\sec^2\theta\tan\theta \text{ mm/s}^2$$ **Ans.**

12 - 25. The cylindrical can C is held fixed while the rod AB and bearings E and F rotate about the vertical axis of the cam at a constant rate of $\dot{\theta} = 4$ rad/s. If the rod is free to slide through the bearings, determine the magnitudes of the velocity and acceleration of the guide D on the rod as a function of θ. The guide follows the groove in the cam, and the groove is defined by the equations $r = 0.25$ m and $z = (0.25\cos\theta)$ m.

Solution

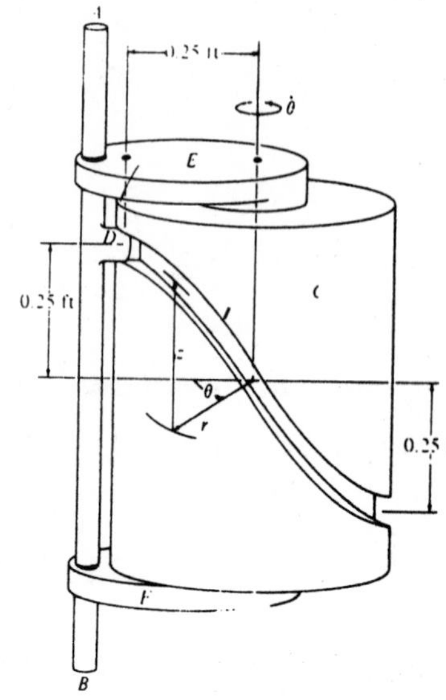

$r =$ ____________________

$\dot{r} =$ ____________________

$\ddot{r} =$ ____________________

$\dot{\theta} =$ ____________________

$\ddot{\theta} =$ ____________________

$z =$ ____________________

$\dot{z} =$ ____________________ $= -\sin\theta$

$\ddot{z} =$ ____________________ $= -4\cos\theta$

$v_r = \dot{r} = 0$

$v_\theta = r\dot{\theta} = (0.25)(4) = 1$

$v_z = \dot{z} = -\sin\theta$

$v = \sqrt{1 + \sin^2\theta}\ \text{m/s}$ **Ans.**

$a_r = \ddot{r} - r\dot{\theta}^2 = 0 - 0.25(4)^2 = -4\ \text{m/s}^2$

$a_\theta = r\ddot{\theta} + 2\dot{r}\dot{\theta} = 0 + 0 = 0$

$a_z = \ddot{z} = -4\cos\theta$

$a = (4\sqrt{1 + \cos^2\theta})\ \text{m/s}^2$ **Ans.**

Absolute Dependent Motion Analysis

12 - 26. If the end of the cable at A is pulled down with a speed of 3 m/s, determine the speed at which block B rises.

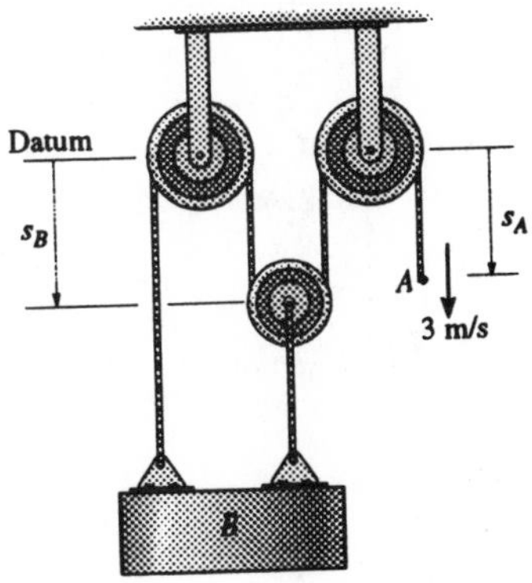

Solution

Use the datum which passes through the fixed points. Measure the position of B using s_B.

Position Equation

____

Velocity Equation

____

$v_B = -1 \text{ m/s} = 1 \text{ m/s} \uparrow$ **Ans.**

12 - 27. If the end of the cable at A is pulled upwards with a speed of 8 m/s, determine the speed at which block B rises.

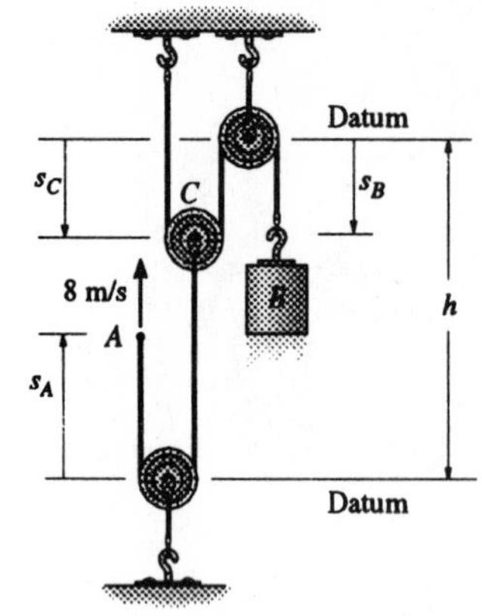

Solution

Here there are two cords and so two position equations must be written, one for each cord. Note that the datums are located through fixed points and that h is a fixed distance.

Position equations

________________________________ $= l$

________________________________ $= l'$

Velocity equations

____

____

When $v_A = 8$ m/s,

$v_C = 8$ m/s

$v_B = -16$ m/s $= 16$ m/s $\uparrow$ **Ans.**

12 - 28. The cable is attached to the block B at E, wraps around three pulleys, and is tied to the back of a truck. I the truck starts from rest when x_D is zero, and moves forward with a constant acceleration of $a_D = 2\ \text{m/s}^2$, determine the speed of block B at the instant $x_D = 3\text{m}$. Neglect the size of the pulleys in the calculation. When $x_D = 0,\ x_B = 5$ m so that points C and D are at the same elevation.

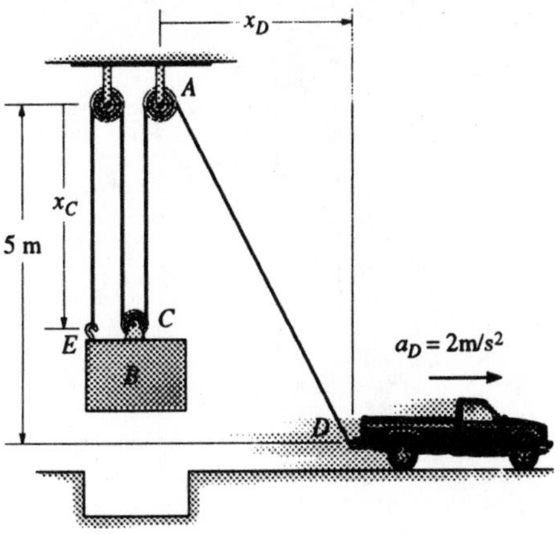

Solution

We must first relate the coordinates x_B and x_D using the problem geometry.
When $x_D = 0$, the total cord length is

$l =$ ______________________

When $x_D > 0$, express the cord length in terms of x_D and x_B.

$l =$ ______________________

Take the time derivative, using the chain rule of calculus.

____ (1)

At $x_D = 3$ m,

$$(\overset{+}{\rightarrow})\ v_D^2 = (v_D)_0^2 + 2a_D[(x_D) - (x_D)_0]$$

$v_D^2 =$ ______________________

$$v_D = \dot{x}_D = 3.46\ \text{m/s}$$

Equation (1) becomes

$$3v_B + \frac{1}{2}[(3)^2 + (5)^2]^{-1/2}(2)(3)(3.46) = 0$$

$v_B = -0.594\ \text{m/s} = 0.594\ \text{m/s} \uparrow$ **Ans.**

12 - 29. If the hoist H is moving upwards at 6 m/s, determine the speed at which the motor M must draw in the supporting cable.

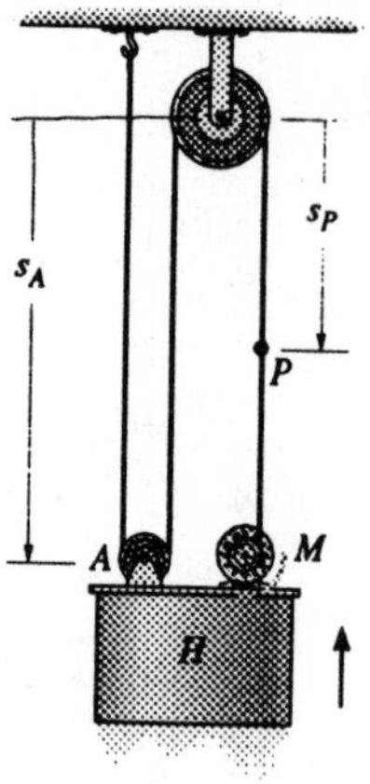

Solution

Establish the datum through top pulley. Specify the position of P and pulley A on hoist H.

Position equation

____

Velocity equation

____

$v_P = 12 \text{ m/s} \downarrow$

Apply the relative velocity equation

$v_P = v_H + v_{P/H}$

____

$v_{P/H} = 18 \text{ m/s} \downarrow$ **Ans.**

Relative Motion Analysis Using Translating Axes

12 - 30. At the instant shown, cars A and B are traveling at speeds of 20 m/s and 45 m/s, respectively. If B is accelerating at 4 m/s^2 while A maintains a constant speed, determine the velocity and acceleration of A with respect to B.

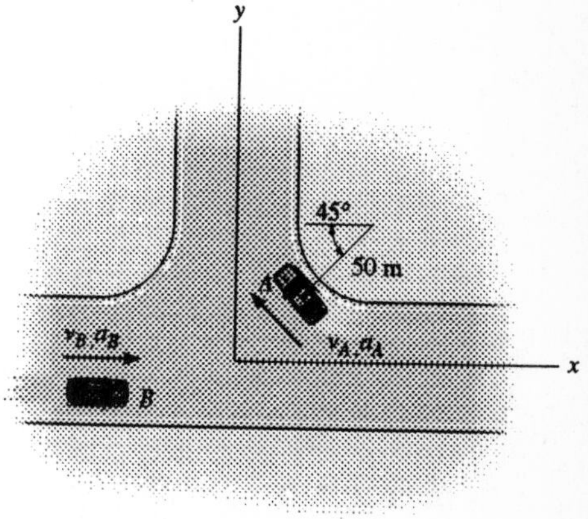

Solution

Assume $\mathbf{v}_{A/B} = (v_{A/B})_x\mathbf{i} + (v_{A/B})_y\mathbf{j}$ and apply the relative velocity equation.

$$\mathbf{v}_A = \mathbf{v}_B + \mathbf{v}_{A/B}$$

____

$(\overset{+}{\rightarrow})\ -20\cos 45° =$ ______________________________

$(+\uparrow)\ \ 20\sin 45° =$ ______________________________

$$(v_{A/B})_x = -59.14\ \text{m/s} = 59.14\ \text{m/s} \leftarrow$$

$$(v_{A/B})_y = 14.14\ \text{m/s} \uparrow$$

$$v_{A/B} = \sqrt{(59.14)^2 + (14.14)^2} = 60.8\ \text{m/s}$$ **Ans.**

$$\theta = \tan^{-1}\frac{14.14}{59.14} = 13.4°$$ **Ans.**

Assume $\mathbf{a}_{A/B} = (a_{A/B})_x\mathbf{i} + (a_{A/B})_y\mathbf{j}$ and apply the relative acceleration equation.

$$\mathbf{a}_A = \mathbf{a}_B + \mathbf{a}_{A/B}$$

____

$(\overset{+}{\rightarrow})\ \ 5.657 = 4 + (a_{A/B})_x$

$(+\uparrow)\ \ 5.657 = 0 + (a_{A/B})_y$

$$(a_{A/B})_x = 1.657\ \text{m/s}^2 \leftarrow$$

$$(a_{A/B})_y = 5.657\ \text{m/s}^2 \uparrow$$

$$a_{A/B} = \sqrt{(1.657)^2 + (5.657)^2} = 5.89\ \text{m/s}^2$$ **Ans**

$$\phi = \tan^{-1}\left(\frac{5.657}{1.657}\right) = 73.7°$$ **Ans.**

12 - 31. The driver of car B observes the motion of the car at A. At the instant shown, car A has a speed of 18 m/s and is reducing its speed at the rate of 1.5 m/s^2. Car B is accelerating at 2 m/s^2 and has a speed of 25 m/s. Determine the magnitudes of the velocity and acceleration of A with respect to B. Car B is moving along a curve having a radius of $r = 300$ m.

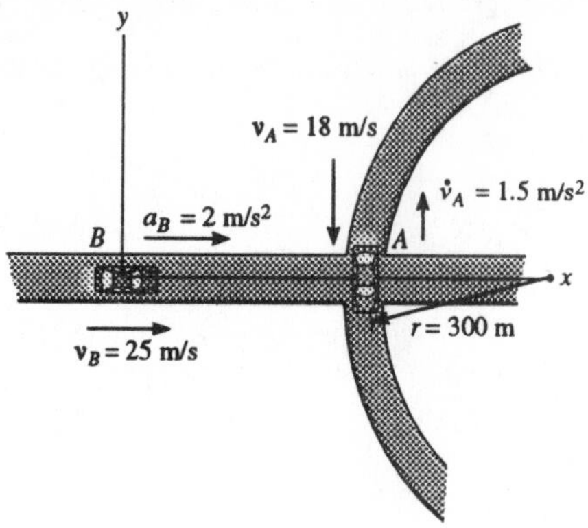

Solution

Assume $\mathbf{v}_{A/B} = (v_{A/B})_x\mathbf{i} + (v_{A/B})_y\mathbf{j}$ and apply the relative velocity equation.

$$\mathbf{v}_A = \mathbf{v}_B + \mathbf{v}_{A/B}$$

$$(v_{A/B})_x = -25 \text{ m/s} = 25 \text{ m/s} \leftarrow$$

$$(v_{A/B})_y = -18 \text{ m/s} = 18 \text{ m/s} \downarrow$$

$$v_{A/B} = \sqrt{(-25)^2 + (-18)^2}$$

$$v_{A/B} = 30.8 \text{ m/s}$$ **Ans.**

Assume $a_{A/B} = (a_{A/B})_x\mathbf{i} + (a_{A/B})_y\mathbf{j}$ and apply the relative acceleration equation

$$\mathbf{a}_A = \mathbf{a}_B + \mathbf{a}_{A/B}$$

$$(a_{A/B})_x = -0.920 \text{ m/s}^2 = 0.920 \text{ m/s}^2 \leftarrow$$

$$(a_{A/B})_y = 1.50 \text{ m/s}^2 \uparrow$$

$$a_{A/B} = \sqrt{(-0.920)^2 + (1.50)^2}$$

$$a_{A/B} = 1.76 \text{ m/s}^2$$ **Ans.**

13 Kinetics of a Particle:

Equations of Motion: Rectangular Coordinates

13 - 1. Each of the three barges has a mass of 30 Mg, whereas the tugboat has a mass of 12 Mg. As the barges are being pulled forward with a constant velocity of 4 m/s, the tugboat must overcome the frictional resistance of the water, which is 2 kN for each barge and 1.5 kN for the tugboat. If the cable between A and B breaks, determine the acceleration of the tugboat.

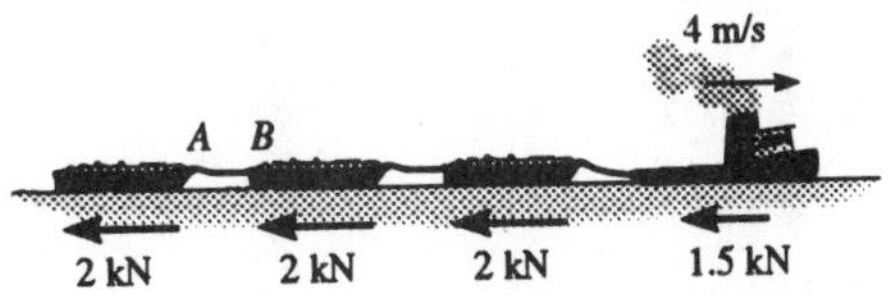

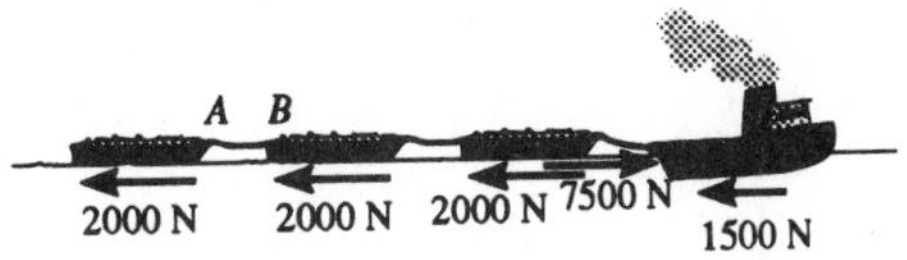

Solution

At constant velocity (equilibrium) the propeller on the tug must exert a force of 7500 N to overcome the water resistance. From a free - body diagram of the system, after the cord breaks, we have

$\overset{+}{\rightarrow} \Sigma F_x = ma_x;$ ______________________________

$a = 0.0278 \text{ m/s}^2 \rightarrow$ **Ans.**

13 - 2. A 2 - kg block is placed on a spring scale located in an elevator that is moving downward. If the scale reading, which measures the force in the spring, is 20 N, determine the acceleration of the elevator. Neglect the mass of the scale.

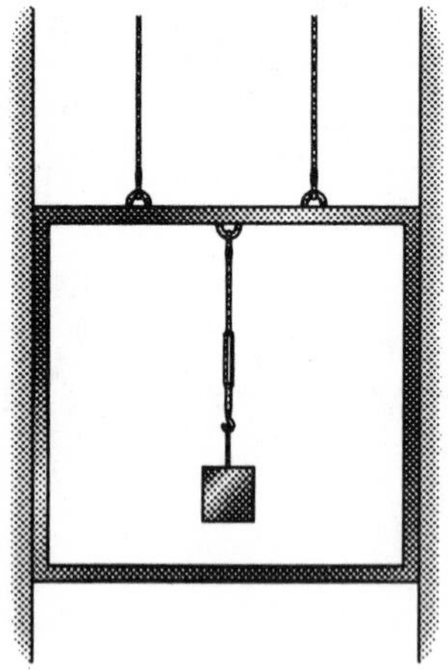

Solution

First we draw the force - body diagram of the block.

$+\downarrow \Sigma F_y = ma_y$;____

$a = -0.190 \text{ m/s}^2 = 0.190 \text{ m/s}^2 \uparrow$ **Ans.**

13 - 3. The 300 - kg bar B, originally at rest, is being towed over a series of small rollers. Compute the force in the cable when $t = 5$ s, if the motor M is drawing in the cable for a short time at a rate of $v = (0.4t^2)$ m/s, where t is in seconds ($0 \leq t \leq 6$s). How far does the bar move in 5 s? Neglect the mass of the cable, pulley P, and the rollers.

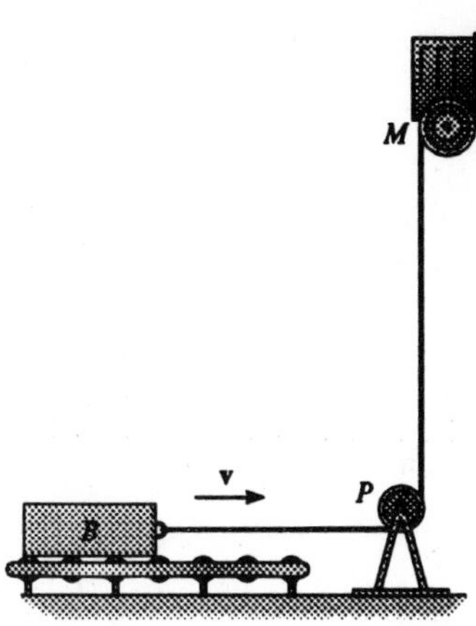

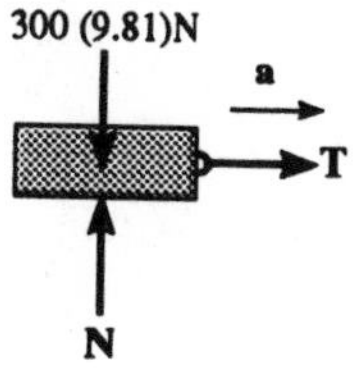

Solution

Since $v = 0.4t^2$; then a as a function of time is

$a =$ ____________________________

When $t = 5$ s,

$a = 4 \text{ m/s}^2$

Using a free - body diagram of the bar,

$\overset{+}{\rightarrow} \Sigma F_x = ma_x$; ____________________________

$T = 1200$ N **Ans.**

Apply $ds = vdt$ to set up the definite integrals for finding s.

____

$$s = \frac{0.4}{3} t^3 \Big|_0^5$$

$s = 16.7$ m **Ans.**

13 - 4. A 2 - kg block is released from rest at A and slides down the inclined plane. If the coefficient of kinetic friction between the plane and the block is $\mu_k = 0.3$, determine the speed of the block after it slides 3 m down the plane.

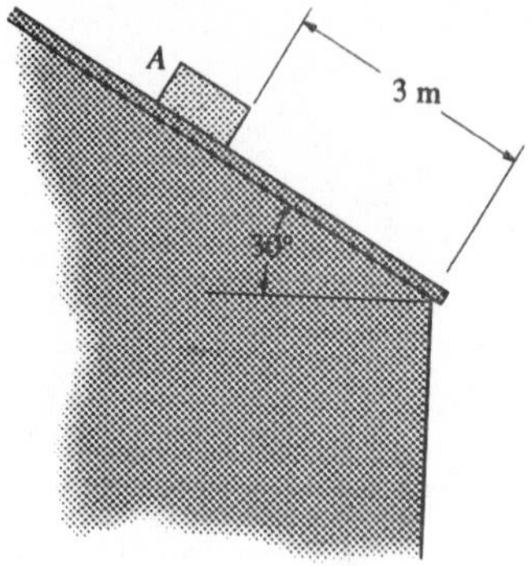

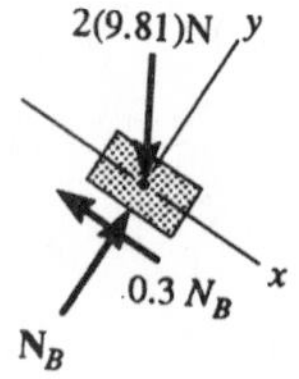

Solution

First we must draw the free - body diagram of the block.

$\searrow + \ \Sigma F_y = ma_y$;____

$N_B = 16.99$ N

$\overset{+}{\nearrow} \Sigma F_x = ma_x$;____

$a = 2.356$ m/s^2

Since the acceleration is constant, then

$\overset{+}{\searrow} \ v_1^2 = v_0^2 + 2a_c(s_1 - s_0)$;____

$v_2 = 3.76$ m/s **Ans.**

13 - 5. The 4 - kg shaft passes through a smooth journal bearing. Initially the spring is unstretched when no force is applied to the shaft. In this position $s = 0.1$ m and the shaft is originally at rest. If a horizontal force of $F = 3$ kN is applied, determine the speed of the shaft at the instant $s = 0.05$ m.

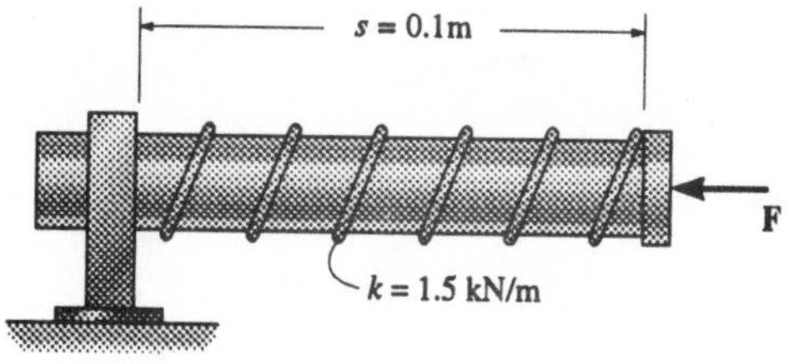

Solution

The free - body diagram is shown only with the horizontal forces acting on the shaft, when the shaft is in the general position.

$\overset{+}{\rightarrow} \Sigma F_x = ma_x$;______________________________

$$a = 750 - 375x$$

Here the acceleration is a function of position. Use kenimatics and set up the definite integrals for finding v when $x = 50$ mm.

____

$$750(0.05) - 187.5(0.05)^2 = \frac{1}{2}v^2$$

$$v = 8.61 \text{ m/s}$$ **Ans.**

13 - 6. Determine the acceleration of block A when the system is released. The surface upon which B rests is smooth. Neglect the mass of the pulleys and cords.

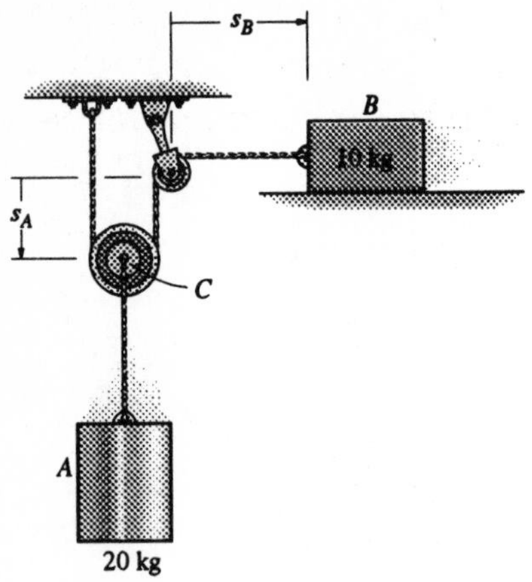

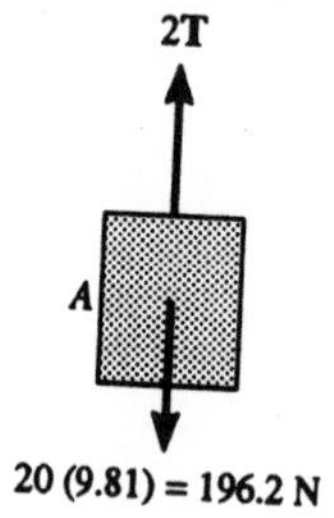

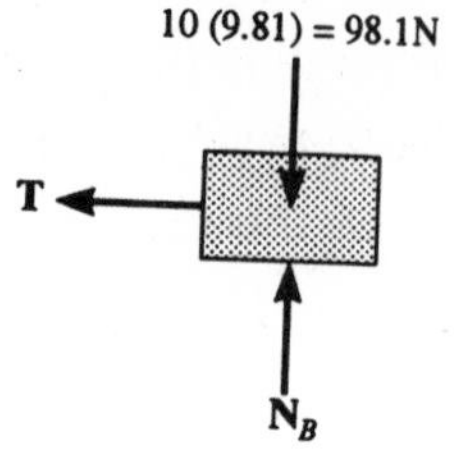

Solution

The free - body diagrams of each block are drawn first.

Block A :

$$+\downarrow \Sigma F_y = ma_y ;\_\_\_\_\_\_\_\_\_\_\_\_\_\_\_\_\_\_\_\_\_\_\_\_\_\_\_\_\_\_\_\_\_\_\_\_\_\_\_\_$$

Block B :

$$\overset{+}{\rightarrow} \Sigma F_x = ma_x ;\_\_\_\_\_\_\_\_\_\_\_\_\_\_\_\_\_\_\_\_\_\_\_\_\_\_\_\_\_\_\_\_\_\_\_\_\_\_\_\_$$

Write the position equation using the datums indicated in the figure.

____

$$2a_A = -a_B$$

Note that the position coordinates for s_A and s_B, and a_A and a_B must be in the *same* directions as indicated. Solving,

$$a_A = 3.27 \text{ m/s}^2 \qquad \textbf{Ans.}$$

$$a_B = -6.54 \text{ m/s}^2 = 6.54 \text{ m/s}^2 \leftarrow$$

$$T = 65.4 \text{ N}$$

Equations of Motion: Normal and Tangential Components

13 - 7. A boy twirls a 2 - kg bucket of water in a vertical circle. If the radius of curvature of the path is 2m, determine the minimum speed the bucket must have when it is overhead at A so no water spills out. Neglect the size of the bucket in the calculation. If the bucket were moving at a slightly slower rate than that calculated, would the water fall on the boy when it starts to spill out at A? Explain.

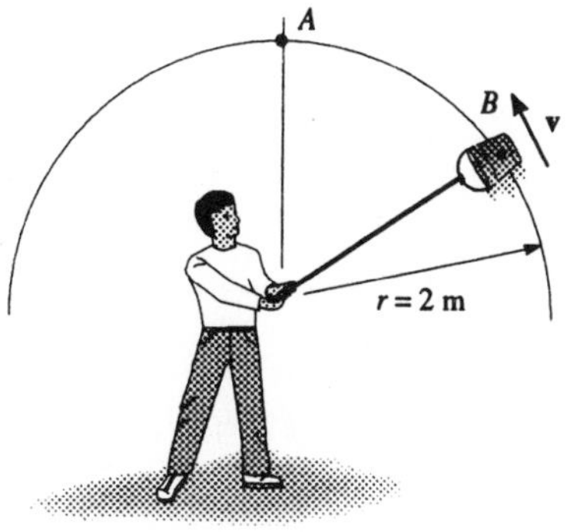

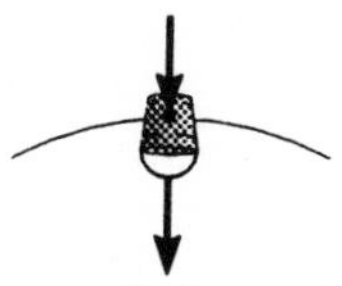

Solution

First the free - body diagram of the bucket is drawn at A. We require $T = 0$ for v_{min}.

$+\downarrow \Sigma F_n = ma_n$;____ .

$v = 4.43$ m/s **Ans.**

Will the water spill on the boy?

No/Yes

Explain____

13 - 8. A toboggan and rider have a total mass of 100 kg and travel down along the (smooth) slope defined by the equation $y = 0.2x^2$. At the instant $x = 8$ m, the tobaggan's speed is 4 m/s. At this point, determine the rate of increase in speed and the normal force which the toboggan exerts on the slope. Neglect the size of the toboggan and rider for the calculation.

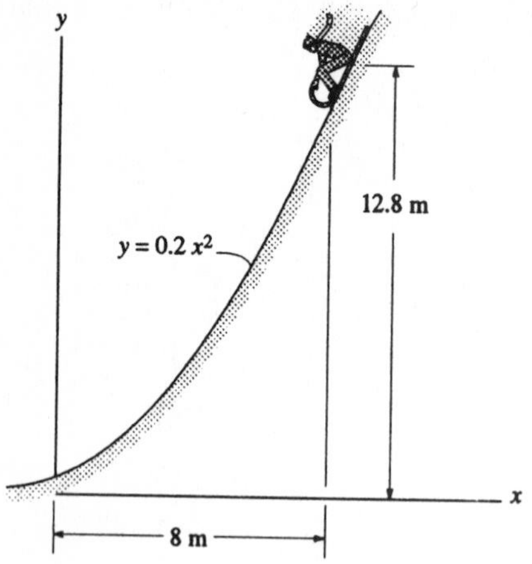

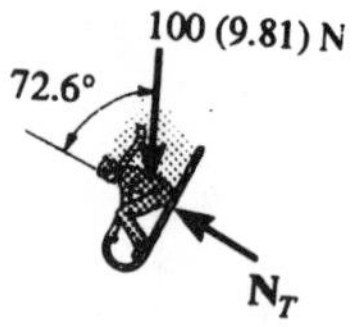

Solution

$$y = 0.2x^2$$

$$\left.\frac{dy}{dx}\right|_{x=8\text{ m}} = 0.4x = 3.2, \quad \theta = \rule{4cm}{0.4pt} = 72.6°$$

$$\frac{d^2y}{dx^2} = 0.4$$

$$\rho = \left|\frac{[1+(\frac{dy}{dx})^2]^{3/2}}{(\frac{d^2y}{dx^2})}\right| = \left|\frac{[1+(3.2)^2]^{3/2}}{0.4}\right| = 94.2 \text{ m}$$

$\nwarrow + \Sigma F_n = ma_n;$ ____

$$N_T = 310 \text{ N}$$ **Ans.**

$\swarrow + \Sigma F_t = ma_t;$ ____

$$a_t = 9.36 \text{ m/s}^2$$ **Ans.**

13-9. A 2-kg block slides on the smooth circular path. If it is released from rest when $\theta = 0°$, determine the force it exerts on the path when it arrives at point A at $\theta = 30°$.

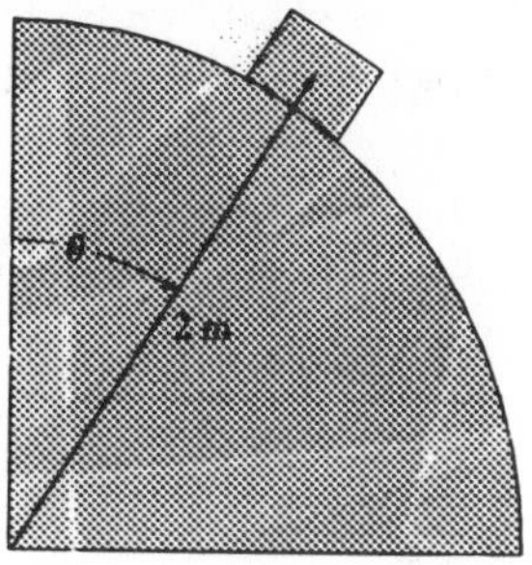

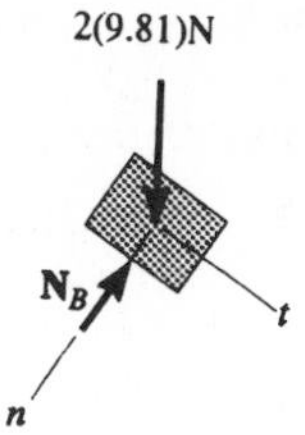

Solution

The free-body diagram of the block is drawn when the block is in the general position θ.

$+\searrow \Sigma F_t = ma_t$;____

$$a_t = 9.81\sin\theta \qquad (1)$$

$+\swarrow \Sigma F_n = ma_n$;____(2)

Here a_t is a function of position θ. Since $ds = 2d\theta$, use kinematics to relate v to θ. Set up the definite integrals.

$$vdv = a_t ds$$

____

Using Eq. (1) :

$$\frac{1}{2}v_A^2 = -19.62[\cos 30° - \cos 0°]$$

$$v_A = 2.29 \text{ m/s}$$

Using Eq. (2) :

At A, $\theta = 30°$

$$N_B - 2(9.8)\cos 30° = -2(\frac{(2.29)^2}{2})$$

$$N_B = 11.7 \text{ N} \qquad \textbf{Ans.}$$

Equations of Motion: Cylindrical Coordinates

13 - 10. The 2 - kg cylinder is moving on the circular path due to the rotation of the forked rod. If the rod maintains a constant angular motion of $\dot{\theta} = 0.5$ rad/s, determine the force which the rod exerts on the cylinder at the instant $\theta = 30°$. Neglect the effects of friction in the calculation.

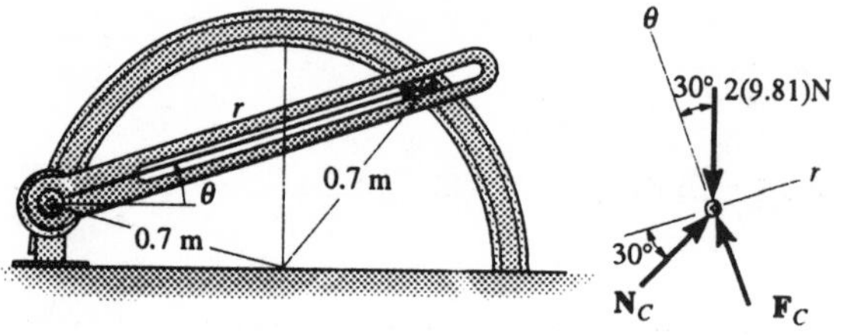

Solution

The free - body diagram is drawn first.

$\nearrow + \Sigma F_r = ma_r;$ ______________________________ $= 2a_r$ (1)

$\nwarrow + \Sigma F_\theta = ma_\theta;$ ______________________________ $= 2a_\theta$ (2)

$\dot{\theta} = 0.5$

$\ddot{\theta} =$ ______________

$r = 2(0.7)\cos\theta = 1.4\cos\theta$

$\dot{r} =$ ______________ $= -0.7\sin\theta$

$\ddot{r} =$ ______________ $= -0.35\cos\theta$

At $\theta = 30°$,

$$a_r = \ddot{r} - r(\dot{\theta})^2 = -0.35\cos 30° - (1.4\cos 30°)(0.5)^2 = -0.6062 \text{ m/s}^2$$

$$a_\theta = r\ddot{\theta} + 2\dot{r}\dot{\theta} = 0 + 2(-0.7\sin 30°)(0.5) = -0.350 \text{ m/s}^2$$

Substituting into Eqs. (1) and (2) and solving

$F_C = 11.3$ N **Ans.**

$N_C = 9.93$ N

13-11. The 2-kg collar slides along the smooth *horizontal* spiral rod, $r = (2\theta)$ m, where θ is in radians. If its angular rate of rotation is constant and equals $\dot{\theta} = 3$ rad/s, determine the horizontal force **P** needed to cause the motion and the normal force that the collar exerts on the rod at the instant $\theta = 90°$.

Solution

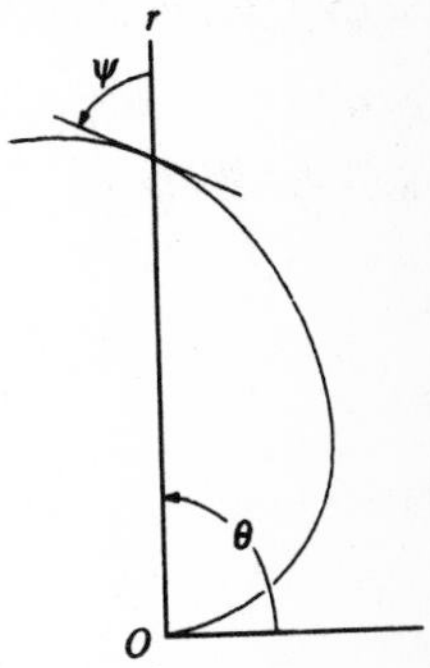

$\tan\psi = r/(dr/d\theta) =$ ________________ $|_{\pi/2} = \pi/2$

$\psi = 57.52°$

The free-body diagram is drawn as shown. We have

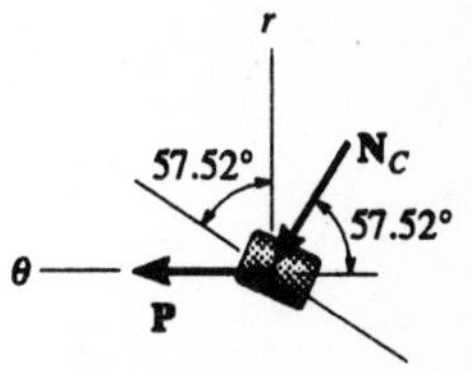

$+\uparrow \Sigma F_r = ma_r;$ ________________ $= 2a_r$ (1)

$\overset{+}{\leftarrow} \Sigma F_\theta = ma_\theta;$ ________________ $= 2a_\theta$ (2)

At $\theta = \pi/2$

$\dot{\theta} =$ ________________

$\ddot{\theta} =$ ________________

$r =$ ________________

$\dot{r} =$ ________________

$\ddot{r} =$ ________________

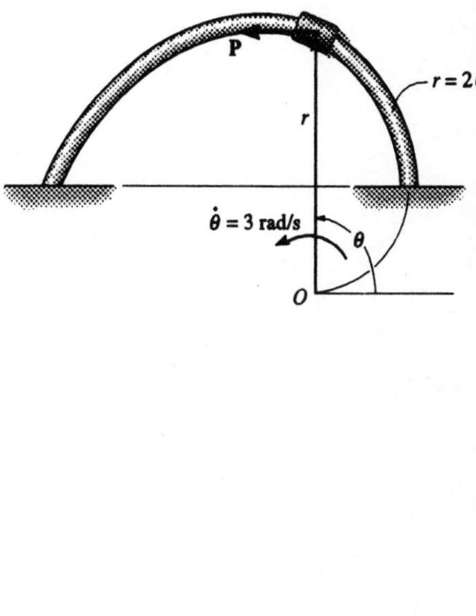

$$a_r = \ddot{r} - r\dot{\theta}^2 = 0 - \pi(3)^2 = -28.27 \text{ m/s}^2$$

$$a_\theta = r\ddot{\theta} + 2\dot{r}\dot{\theta} = 0 + 2(6)(3) = 36 \text{ m/s}^2$$

Substituting into Eqs. (1) and (2) and solving yields

$P = 36.0$ N **Ans.**

$N_C = 67.0$ N **Ans.**

13-12. Rod OA rotates counterclockwise with a constant angular rate of $\dot{\theta} = 5$ rad/s. The double collar C is pin-connected together such that one collar slides over the rotating rod and the other slides over the *horizontal* curved rod, of which the shape is a limacon described by the equation $r = 0.15(2 - \cos\theta)$ m. If both collars have a mass of 0.75 kg, determine the normal force which the curved path exerts on one of the collars, and the force that OA exerts on the other collar at the instant $\theta = 90°$.

Solution

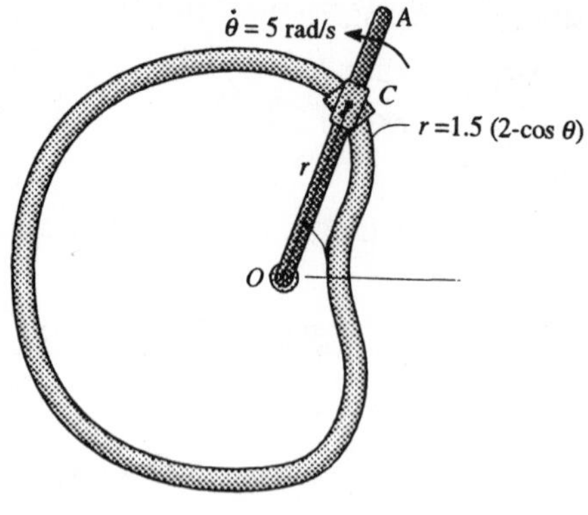

$r = 0.15(2 - \cos\theta)$

$\dfrac{dr}{d\theta} =$ ______________________

At $\theta = 90°$,

$\tan\psi = r/(dr/d\theta) =$ ______________________ $= 2$

$\psi = 63.43°$

The free-body diagram is drawn as shown. We have

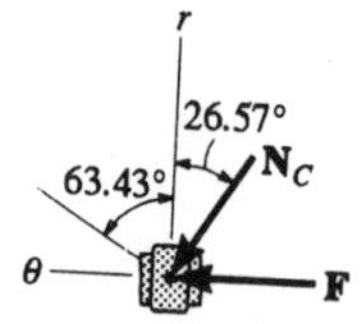

$+\uparrow \Sigma F_r = ma_r;$ ____ $= 0.75a_r$

$\overset{+}{\leftarrow} \Sigma F_\theta = ma_\theta;$ ____ $= 0.75a_\theta$

At $\theta = 90°$

$\dot{\theta} = 5$

$\ddot{\theta} = 0$

$r = 0.15(2 - \cos\theta) = 0.3$ m

$\dot{r} = 0.15(\sin\theta)\dot{\theta} = 0.75$ m/s

$\ddot{r} = 0.15(\cos\theta)\dot{\theta}^2 + 0.15\sin\theta(\ddot{\theta}) = 0$

$a_r = \ddot{r} - r(\dot{\theta})^2 = 0 - 0.3(5)^2 = -7.5 \text{ m/s}^2$

$a_\theta = r\ddot{\theta} + 2\dot{r}\dot{\theta} = 0 + 2(0.75)(5) = 7.5 \text{ m/s}^2$

Subsituting these values into the equations of motion and solving;

$N_C = 6.29$ N **Ans.**

$F = 2.81$ N **Ans.**

14 Kinetics of a Particle: Work and Energy

Principle of Work and Energy

14-1. A car having a mass of 2 Mg strikes a smooth rigid sign post with an initial speed of 30 km/h. To stop the car, the front end horizontally deforms 0.2 m. If the car is free to roll during the collision, determine the *average* horizontal force causing the deformation.

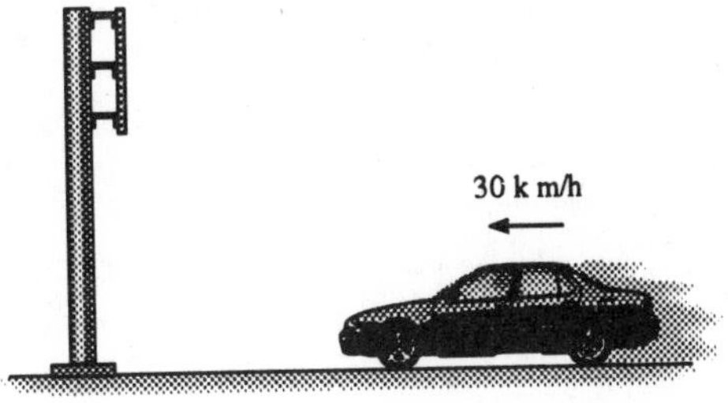

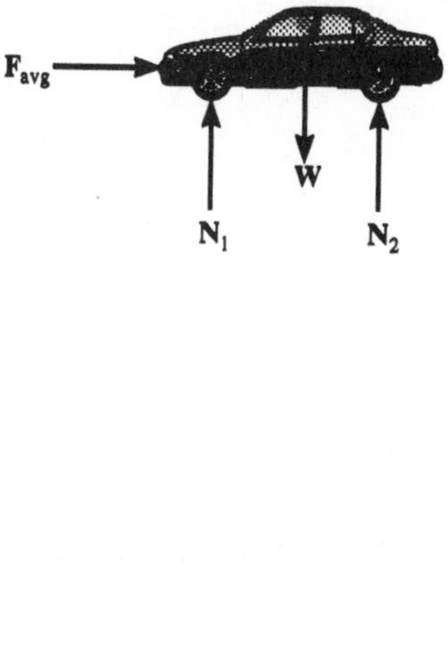

Solution

$v_1 = 30 \text{ km/h} = 8.33 \text{ m/s}$

$\Delta s = 0.20 \text{ m}$

Using the free-body diagram, apply the equation of work and energy.

$T_1 + \Sigma U_{1-2} = T_2$

$F_{avg} = 347 \text{ kN}$ **Ans.**

14 - 2. A car, assumed to be rigid and having a mass of 800 kg, strikes a barrel barrier installation without the driver applying the brakes. From experiments, the magnitude of the force of resistance, $\mathbf{F}_r$, created by deforming the barrels successively, is shown as a function of vehicle penetration. If the car strikes the barrier traveling at $v_c = 70$ km/h, determine approximately the distance s to which the car penetrates the barrier.

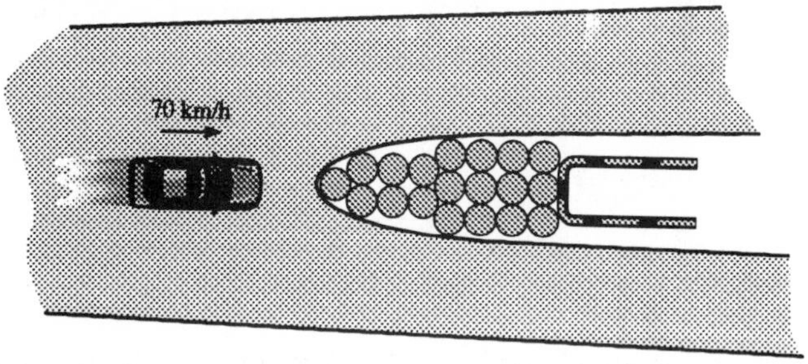

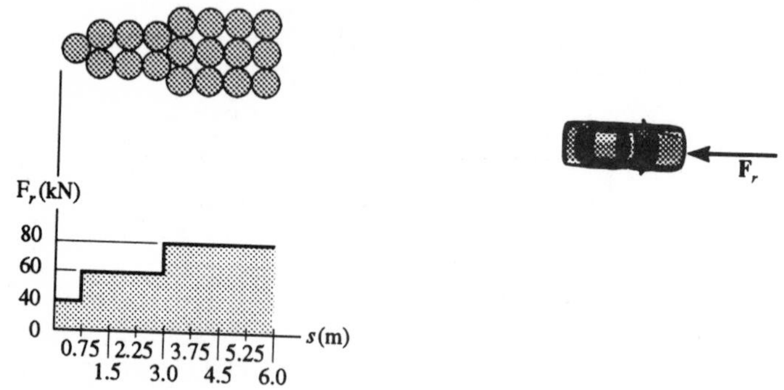

Solution

From the free - body diagram, negative work is done by F_r as the car penetrates the barrier. This work is equal to the area under the $F_r - s$ graph. We require the car's initial kinetic energy to equal this area.

$$T_1 + \Sigma U_{1-2} = T_2$$

$s = 2.77$ m **Ans.**

14 - 3. When a 2 - kg box is at A it has a speed of $v_A = 2$ m/s. Determine its speed at B and the distance x where it stikes the ground at C. The ramp is smooth.

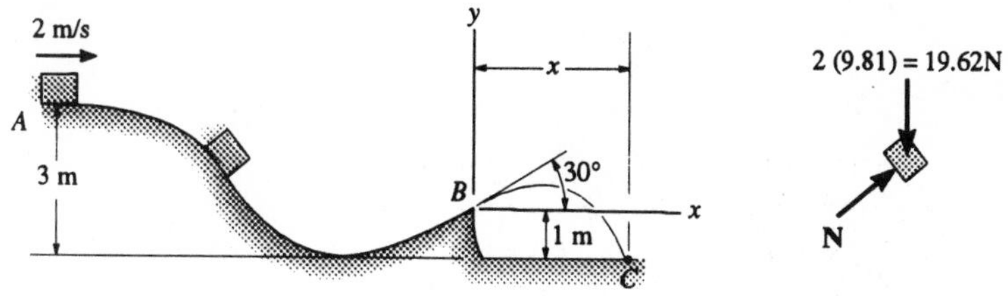

Solution

From the free - body diagram, only the weight does work as the box travels from A to B.

$$T_A + \Sigma U_{AB} = T_B$$

$$v_B = 6.576 \text{ m/s}$$ **Ans.**

$$(v_B)_x = 6.576 \cos 30° = 5.694 \text{ m/s} \rightarrow$$

$$(v_B)_y = 6.576 \sin 30° = 3.288 \text{ m/s} \uparrow$$

Apply the equations of constant acceleration to determine the range R and the time t.

$(\overset{+}{\rightarrow})$ ______________________________

$(+\uparrow)$ ______________________________

Solving for the positive root :

$$t = 0.897 \text{ s}$$

Thus

$$x = (5.694 \text{ m/s})(0.897 \text{ s}) = 5.11 \text{ m}$$ **Ans.**

14 - 4. The coefficient of kinetic friction between the 2 - kg block and the surface is $\mu_k = 0.2$. The block is acted upon by a horizontal force of $P = 40$ N and has a speed of 2 m/s when it is at the point A. Determine the maximum deformation of the outer spring B at the instant the block comes to rest. Spring B has a stiffness of $k_B = 800$ N/m and the "nested" spring C has a stiffnes of $k_C = 400$ N/m.

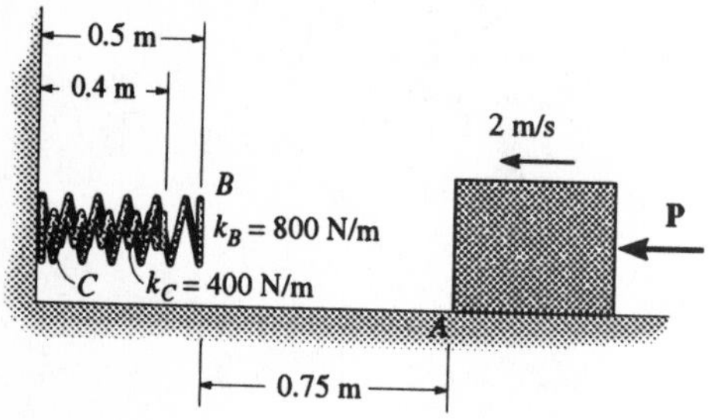

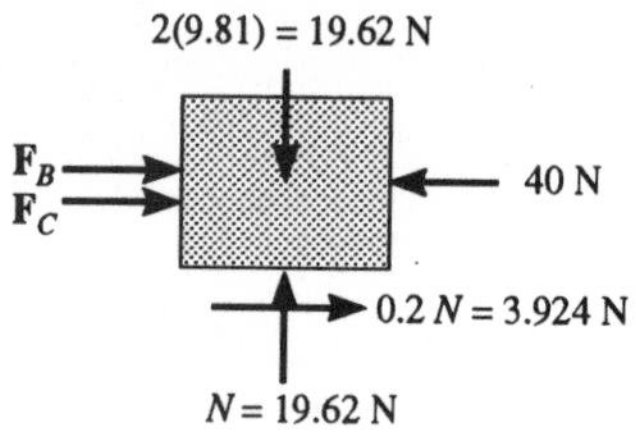

Solution

From the free - body diagram the 40 N force does positive work and friction andthe spring forces do negative work. Write the equation of work and energy assuming spring B is compressed a distance s.

$$T_1 + \Sigma U_{1-2} = T_2$$

$$-600\, s^2 + 76.076\, s + 33.057 = 0$$

Solving for the positive root,

$s = 0.308$ m (compression of B) **Ans.**

$s - 0.1 = 0.208$ m (compression of C)

14 - 5. A motor hoists a 50 - kg crate at constant speed to a height of $h = 6$ m in 3 s. If the indicated power of the motor is 4 kW, determine the motor's efficiency.

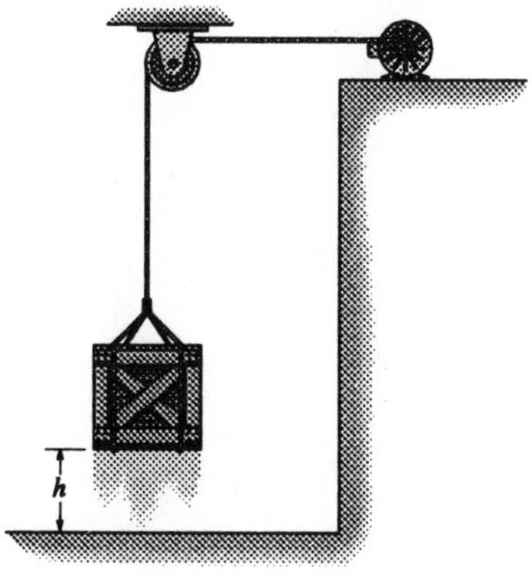

Solution

The motor must overcome the tension in the cable to lift the crate. From the free - body diagram this tension is $T = 490.5$ N since the crate is moving upwards with constant velocity. The work done is

$U_{1-2} =$ ________________________________

$U_{1-2} = 2943$ J

$P_{out} =$ ________________________________

$P_{out} = 981$ W

$\varepsilon =$ ________________________________

$\varepsilon = 0.245$ **Ans.**

14-6. A 2-Mg car climbs a 15° incline at a constant speed of 4 m/s. Determine the power output.

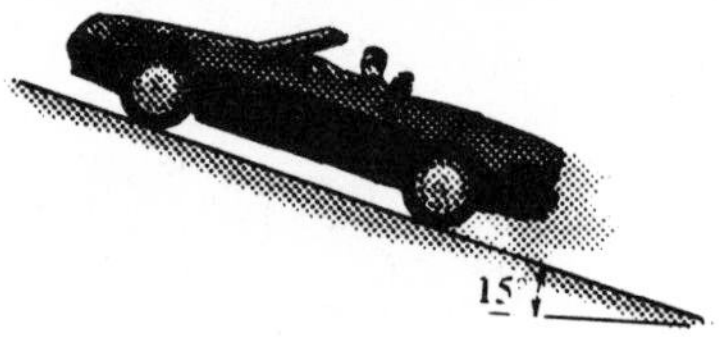

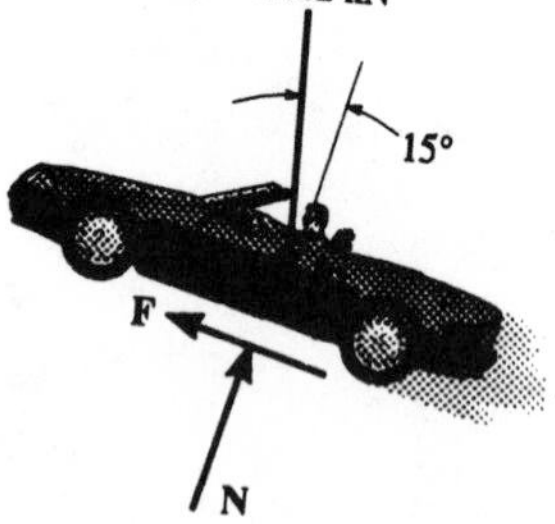

Solution

First we draw the free-body diagram in order to obtain the force doing the work. By inspection, this force is the friction tractive force. Since the car moves with constant velocity,

$$\nwarrow + \Sigma F_x = 0; \; F = \underline{\hspace{5cm}} = 5.078 \text{ kN}$$

$$P = \mathbf{F} \cdot \mathbf{v}; \; \underline{\hspace{8cm}}$$

$$P = 102 \text{ kW}$$ **Ans.**

14 - 7. The motor exerts a force of 400 N on the cable. If the 100 - kg block is originally at rest, determine the power output of the motor in $t = 3$ s.

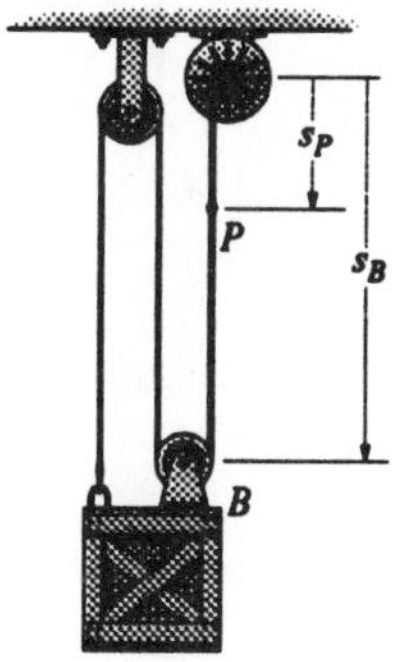

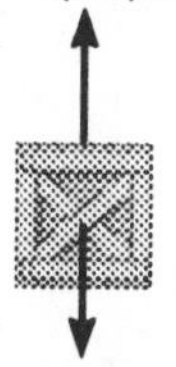

Solution

To solve this problem we must find the speed at which the cable is drawn into the motor. From the free - body diagram of the block, we have

$+\uparrow \Sigma F_y = ma_y$;____

$a = 2.19 \text{ m/s}^2$

Since the acceleration is constant, set up the kinematic equation used to find the velocity of the block when $t = 3$ s.

$(+\uparrow)$____

$v = 6.57$ m/s

Relate the position coordinates s_p and s_B

____

$v_P =$ __________________________________ $= 19.71$ m/s

Thus

$P =$ __________________________________ $= 7.88$ kW **Ans.**

Conservation of Energy Theorem

14 - 8. The 2 - kg block A slides in the smooth horizontal slot. If the block is drawn back so that $s = 3$ m and released from rest, determine its speed at the instant $s = 0$. The spring has a stiffness of $k = 40$ N/m and an unstretched length of 1 m.

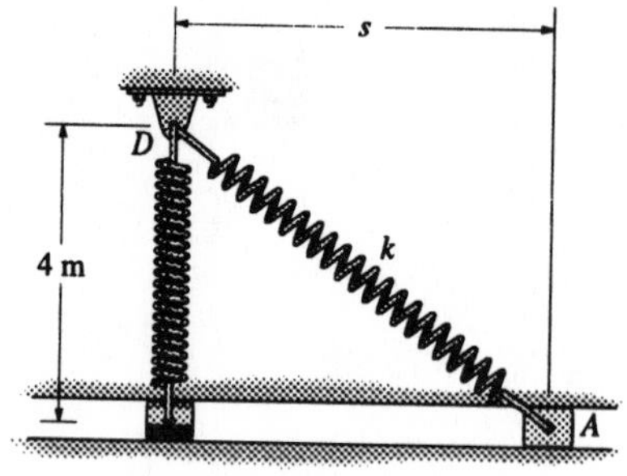

Solution

When the block is at $s = 3$ m, the spring has a length of 5 m and is stretched $(5 - 1)$ m $= 4$ m. When $s = 0$ it is stretched 3 m. The weight remains at the same elevation.

$$T_1 + V_1 = T_2 + V_2$$

$v_2 = 17.3$ m/s **Ans.**

14 - 9. The 50 - kg block A is released from rest. Determine its speed when it has descended 2 m. Block B has a mass of 10 kg and rests on the smooth surface. Neglect the mass of the pulleys.

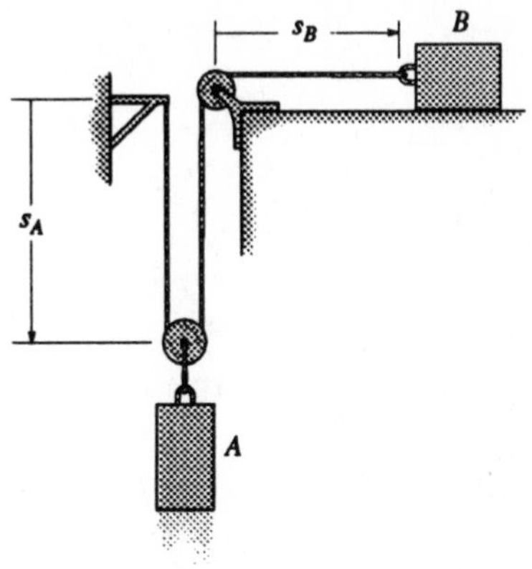

Solution

The speed of block A must be related to the speed of B. Write the position equation for the blocks.

Thus,

$$v_B = -2v_A$$

Use a datum at the initial elevation of block A.

$$T_1 + V_1 = T_2 + V_2$$

$$v_A = 4.67 \text{ m/s}$$ **Ans.**

14 - 10. Two springs are nested together and have the unstretched lengths shown. If a 5 - kg ball is dropped from rest at a height of 5 m onto the springs, determine the maximum deformation of each spring needed to momentarily stop the motion.

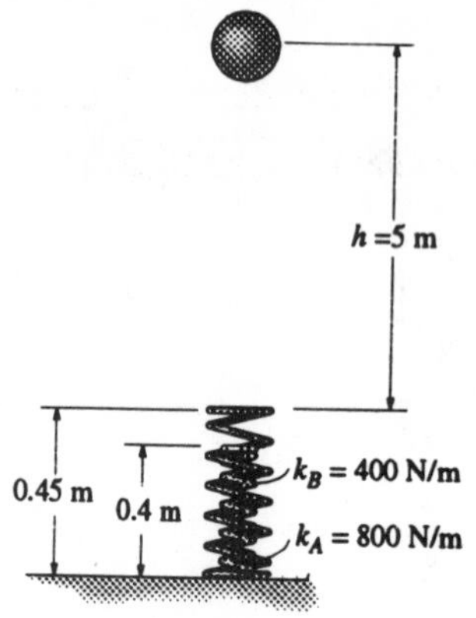

Solution

We will establish the datum at the initial position of the ball and assume both springs deform such that A deforms s_A and B deforms $(s_A - 0.05 \text{ m})$. Thus,

$$T_1 + V_1 = T_2 + V_2$$

$$600 s_A^2 - 69.05 s_A - 19.12 = 0$$

Choosing the positive root,

$$s_A = 0.245 \text{ m}$$

Therefore,

$$s_B = 0.245 \text{ m} - 0.05 \text{ m} = 0.195 \text{ m}$$ **Ans.**

15 Kinetics of a Particle: Impulse and Momentum

Principle of Linear Impulse and Momentum

15 - 1. A hockey puck is traveling to the left with a velocity of $v_1 = 10$ m/s when it is struck by a hockey stick and given a velocity of $v_2 = 20$ m/s as shown. Determine the magnitude of the net impulse exerted by the hockey stick on the puck. The puck has a mass of 0.2 kg.

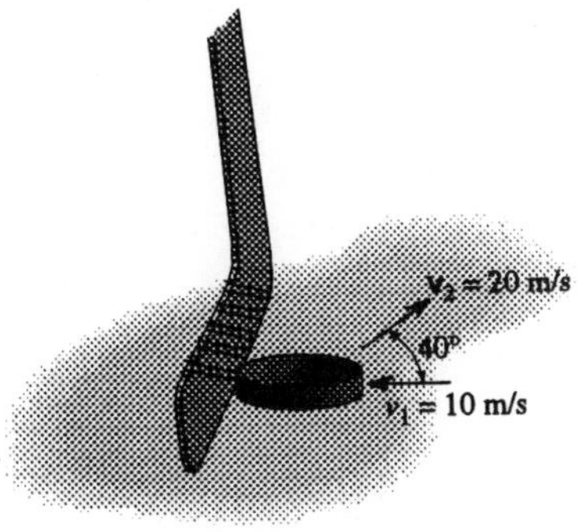

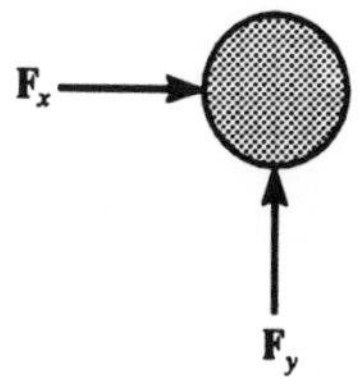

Solution

The free - body diagram shows the x and y components of the impulsive force acting on the puck.

$$(\overset{+}{\rightarrow})\ m(v_x)_1 + \Sigma \int F_x\,dt = m(v_x)_2$$

____

$$\int F_x\,dt = 5.06\ \text{N}\cdot\text{s}$$

$$(+\uparrow)\ m(v_y)_1 + \Sigma \int F_y\,dt = m(v_y)_2$$

____

$$\int F_y\,dt = 2.57\ \text{N}\cdot\text{s}$$

Imp = ______________________________ = 5.68 N · s **Ans.**

15 - 2. A 10 - kg block is initially moving along a smooth horizontal surface with a speed of $v_1 = 3$ m/s. If it is acted upon by a force $F = 30\cos(\frac{\pi}{4}t)$ N, which varies in the manner shown, determine the velocity of the block in 2 s.

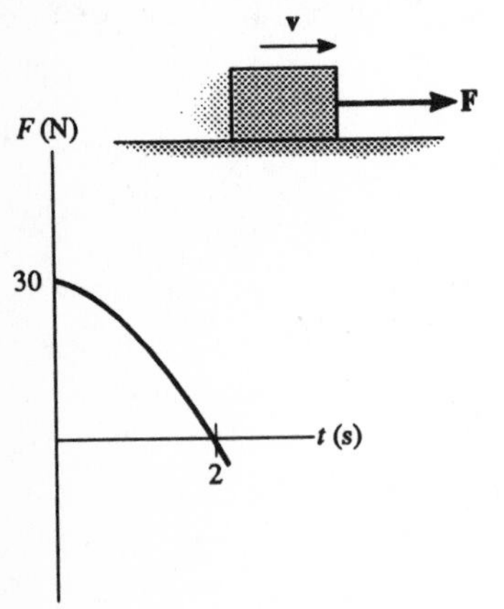

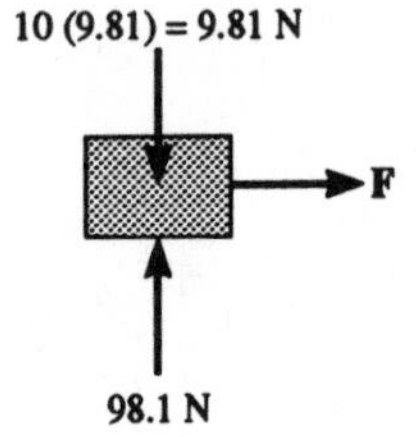

Solution

The free - body diagram is drawn first.

$$(\overset{+}{\rightarrow})\quad m(v_x)_1 + \Sigma\int F_x\,dt = m(v_x)_2$$

$$30 + [30(\frac{4}{\pi})\sin(\frac{\pi}{4}t)\Big|_0^2 = 10v_2$$

$$v_2 = 6.82 \text{ m/s} \rightarrow$$ **Ans.**

15 - 3. The motor M pulls on the cable with a force **F** that has a magnitude which varies as shown on the graph. If the 15 - kg crate is originally resting on the floor determine the speed of the crate in $t = 6$ s after the motor is turned on.

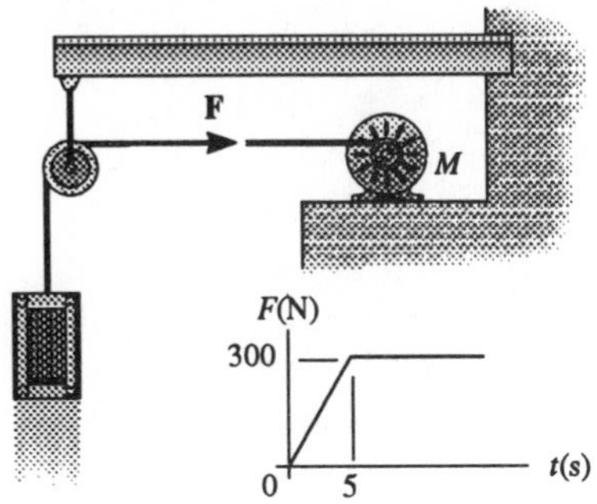

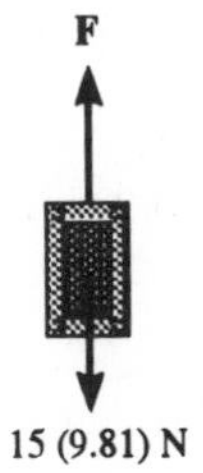

Solution

First we will determine the time needed to lift the crate.
From the free - body diagram, the force needed to lift the crate is

$$F = 15(9.81)\text{ N} = 147.2\text{ N}.$$

From the graph,

$$0 \le t \le 5\text{ s},\ F = \frac{300}{5}t = 60t$$

Hence, the time to lift the crate is

$$t = \underline{\hspace{6cm}} = 2.45\text{ s}$$

The impulse of the motor is calculated as the area under the graph.

$$(+\uparrow)\ m(v_y)_1 + \Sigma\int F_y\,dt = m(v_y)_2$$

$$v_2 = 23.2\text{ m/s}$$ **Ans.**

15 - 4. Block A has a mass of 10 kg and block B has a mass of 5 kg. Determine their velocities in 2 s after they are released from rest. Neglect the mass of the pulleys and cables.

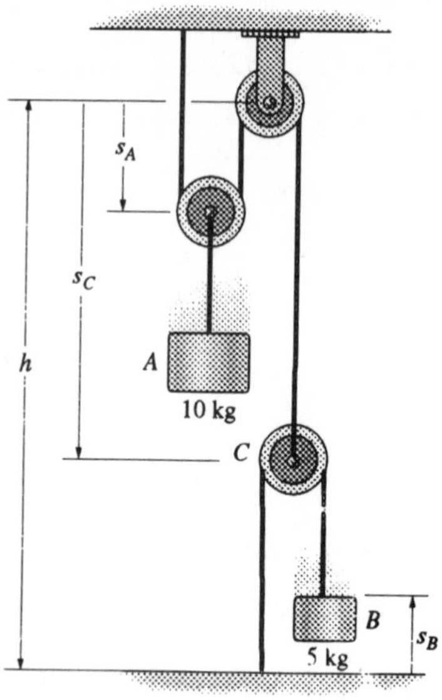

Solution

The free - body diagrams of each block are drawn first.

Block A :

$$(+\downarrow)\quad m(v_A)_1 + \Sigma\int F_y\,dt = m(v_A)_2$$

Block B :

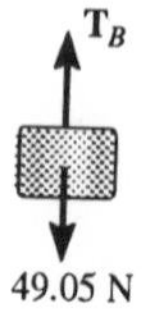

$$(+\downarrow)\quad m(v_B)_1 + \Sigma\int F_y\,dt = m(v_B)_2$$

From the free - body diagrams of the pulleys, relate T_A to T_B.

Using the datums indicated in the figure write the position equations for each cord.

Thus,

$$2v_A = v_B$$

Note that v_A and v_B are *both* assumed to have the same downward directions in *all* corresponding equations. Solving,

$v_A = 4.90\text{ m/s}\downarrow$ **Ans.**

$v_B = -9.81\text{ m/s} = 9.81\text{ m/s}\uparrow$ **Ans**

$T = 73.6\text{ N}$

Conservation of Linear Momentum

15 - 5. A rifle has a mass of 2.5 kg. If it is loosely gripped and a 1.5 - g bullet is fired from it with a muzzle velocity of 1400 m/s, determine the recoil velocity of the rifle just after firing.

Solution

The free - body diagram of the rifle - bullet system indicates that momentum is conserved.

$$(\overset{+}{\rightarrow})\ \Sigma m v_1 = \Sigma m v_2$$

$$(v_R)_2 = 0.840 \text{ m/s} \qquad \textbf{Ans.}$$

15 - 6. A 0.6 - kg brick is thrown into a 25 - kg wagon which is initially at rest. If, upon entering, the brick has a velocity of 10 m/s as shown, determine the final velocity of the brick and wagon.

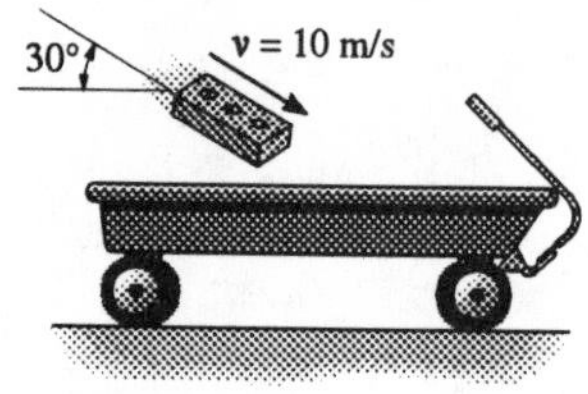

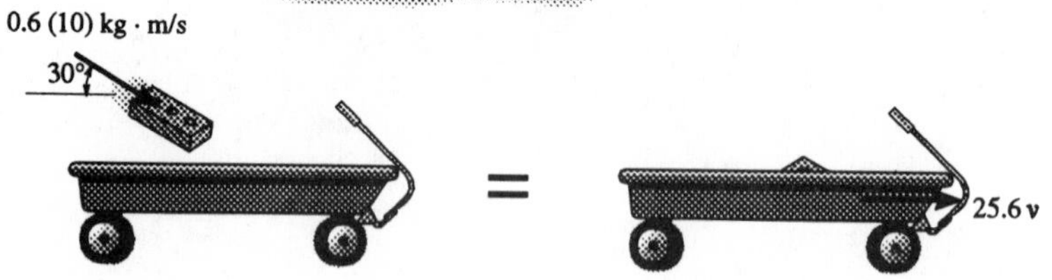

Solution

The impulse of the brick on the wagon and the wagon on the brick are equal and opposite in the horizontal direction and so momentum for this system is conserved in this direction.

$$(\overset{+}{\rightarrow})\quad \Sigma m v_1 = \Sigma m v_2$$

$$v = 0.203 \text{ m/s} \qquad \textbf{Ans.}$$

15 - 7. The two flat cars A and B each have a mass of 80 kg. If the man C has a mass of 70 kg and jumps from A with a horizontal *relative* velocity of $v_{C/A} = 2$ m/s and lands on B, determine the velocity of each car after the jump. Neglect the effects of rolling resistance.

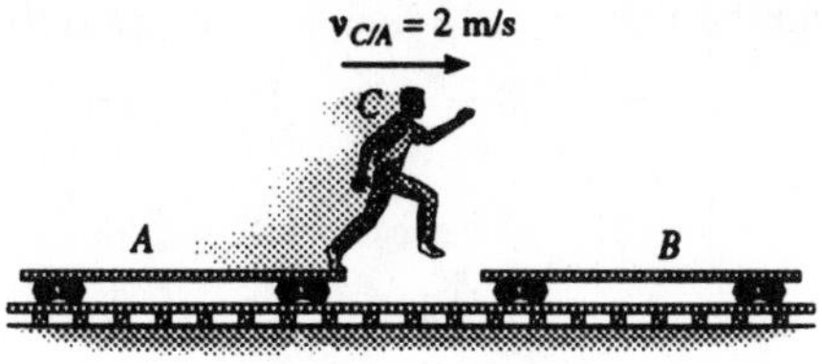

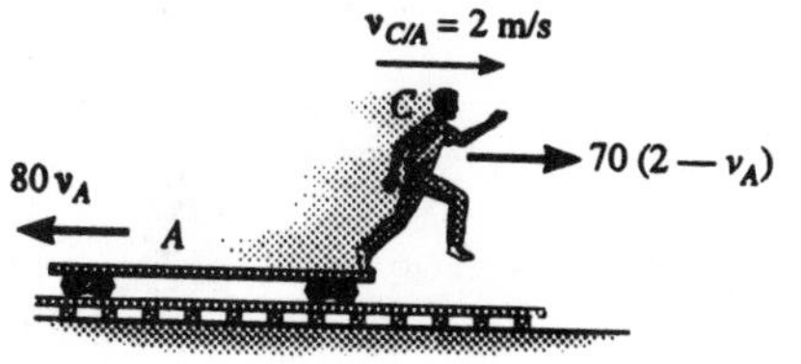

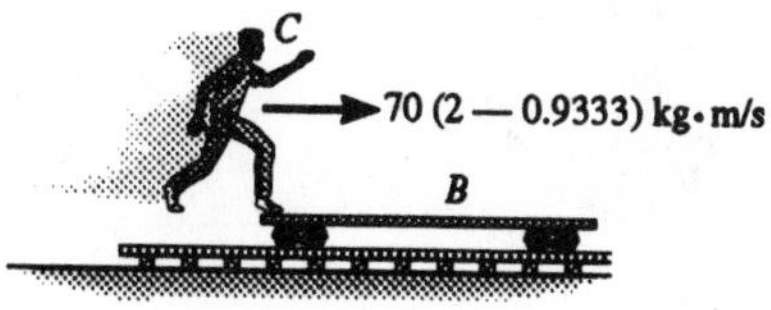

Solution

The horizontal forces between the man and car and car and man are equal but opposite and so momentum for this system is conserved in this direction.

Cart A and man :

$$(\overset{+}{\rightarrow})\ \Sigma m v_1 = \Sigma m v_2$$

$$v_A = 0.933 \text{ m/s} \qquad \textbf{Ans.}$$

Car B and man :

$$(\overset{+}{\rightarrow})\ \Sigma m v_1 = \Sigma m v_2$$

$$v_B = 0.498 \text{ m/s} \qquad \textbf{Ans.}$$

15 - 8. A 70 - kg man wearing ice skates throws an 8 - kg block with an initial velocity of 2 m/s, measued relative to himself, in the direction shown. If he is originally at rest and completes the throw in 1.5 s while keeping his legs rigid, determine the horizontal velocity of the man just after releasing the block. What is the average vertical reaction of both his skates on the ice during the throw? Neglect friction and the motion of his arms.

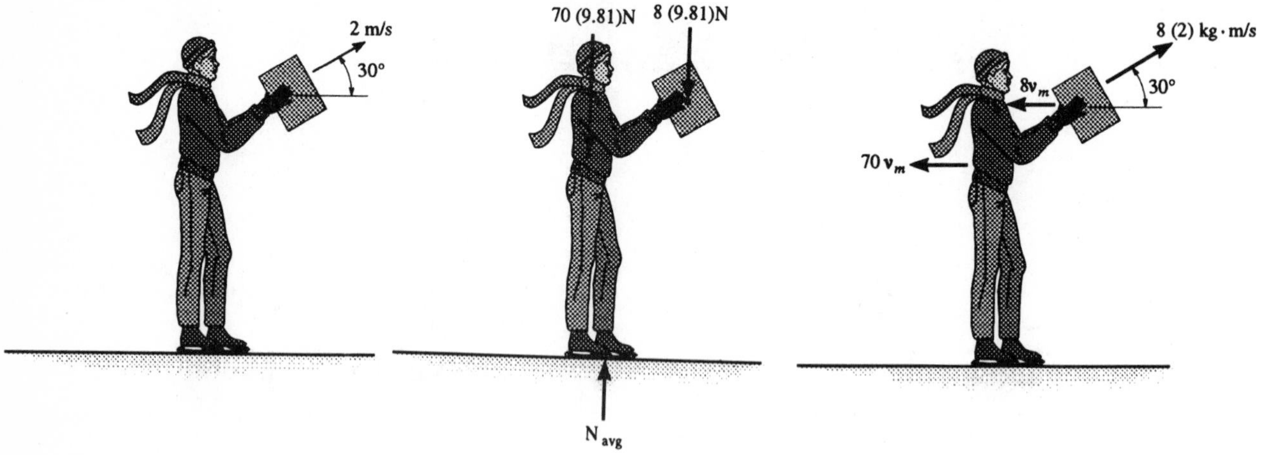

Solution

During the throw the impulses of the man on the block and block on the man are equal but opposite. Momentum is conserved in the x - direction. Assuming the man moves back at v_m, the block will move forward with a velocity of $2\cos 30° - v_m$. Thus

$$(\overset{+}{\rightarrow})\ \Sigma m v_1 = \Sigma m v_2$$

$$v_m = 0.178 \text{ m/s}$$ **Ans.**

$$(+\uparrow)\ \Sigma m v_1 + \Sigma \int F dt = \Sigma m v_2$$

$$N_{avg} = 771 \text{ N}$$ **Ans.**

15-9. Ball B has a mass of 0.5 kg and is moving forward with a velocity of $(v_B)_1 = 2$ m/s when it strikes the 2-kg block A, which is originally at rest. If the coefficient of restitution between the ball and the block is $e = 0.4$, determine (a) the velocity of A and B just after the collision and (b) the distance block A slides before coming to rest. The coefficient of kinetic friction between the block and the surface is $\mu_s = 0.4$.

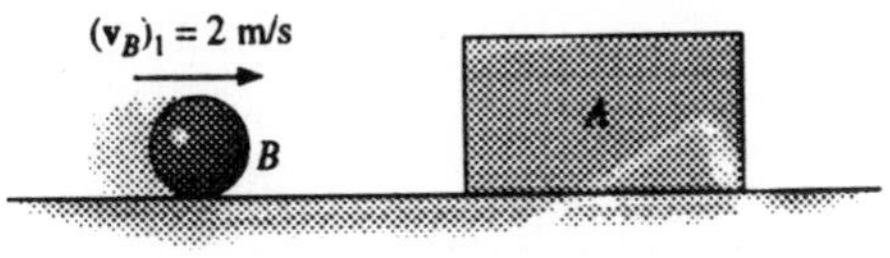

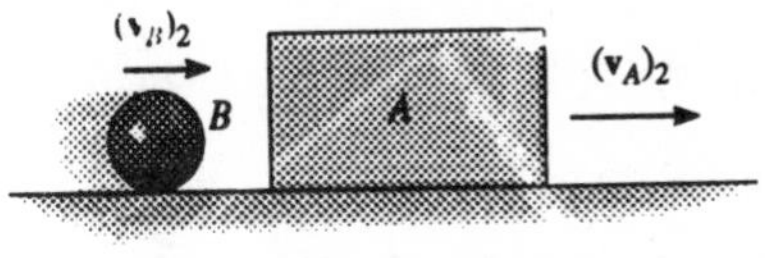

Solution

For the system of the block and ball

$$(\overset{+}{\rightarrow})\quad \Sigma m v_1 = \Sigma m v_2$$

____ (1)

$$(\overset{+}{\rightarrow})\quad e = \frac{(v_A)_2 - (v_B)_2}{(v_B)_1 - (v_A)_1}$$

____ (2)

Solving Eqs. (1) and (2) :

$(v_B)_2 = -0.240 \text{ m/s} = 0.240 \text{ m/s} \leftarrow$ **Ans.**

$(v_A)_2 = 0.560 \text{ m/s} \rightarrow$ **Ans.**

Using the free-body diagram of the block, we have

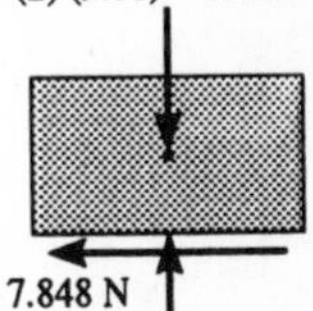

$$T_1 + \Sigma U_{1-2} = T_2$$

____

$x = 40.0$ mm **Ans.**

15-10. A stunt driver in car A travels in free flight off the edge of a ramp at C. At the point of maximum height he strikes car B. If the direct collision is perfectly plastic ($e = 0$), determine the required ramp speed v_C at the end of the ramp C, and the approximate distance s where both cars strike the ground. Each car has a mass of 3.5 Mg. Neglect the size of the cars in the calculation.

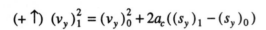

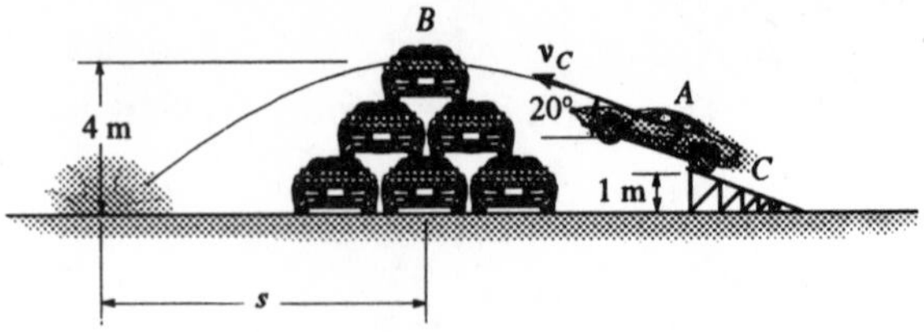

Solution

Determine the velocity the car must have at C.

$$(+\uparrow)\quad (v_y)_1^2 = (v_y)_0^2 + 2a_c((s_y)_1 - (s_y)_0)$$

____

$$v_C = 22.43 \text{ m/s} = 80.8 \text{ km/h}$$ **Ans.**

The velocity just before the collision is

$(v_C)_x =$ ____ $= 21.08$ m/s

The velocity of both cars just after collision is

$$(\overset{+}{\leftarrow})\quad \Sigma m v_1 = \Sigma m v_2$$

____

$$v_2 = 10.54 \text{ m/s} \leftarrow$$

Time to fall is

$$(+\downarrow)\quad (s_y)_2 = (s_y)_1 + (v_y)_1 t + \tfrac{1}{2}a_c t^2$$

____

$$t = 0.903 \text{ s}$$

The distance s is

$$(\overset{+}{\leftarrow})\quad (s_x)_2 = (s_x)_1 + (v_x)_1 t$$

____

$$s = 9.52 \text{ m}$$ **Ans.**

15-11. Plates A and B each have a mass of 4 kg and are restricted to move along the frictionless guides. If the coefficient of restitution between the plates is $e = 0.7$, determine (a) the speed of both plates just after collision and (b) the maximum deflection of the spring. Plate A has a velocity of 4 m/s just before striking B. Plate B is originally at rest.

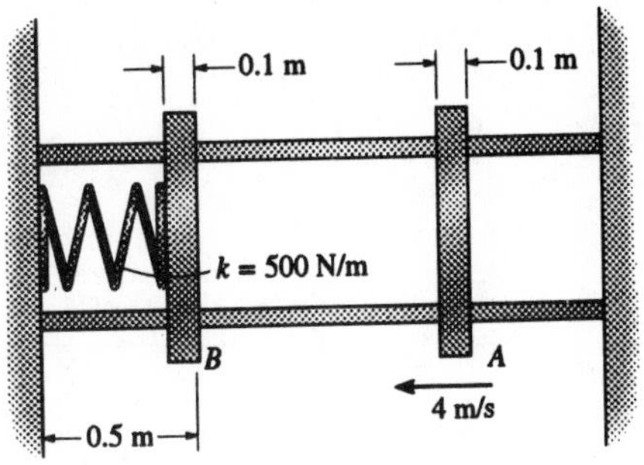

Solution

During the collision plate B deforms the spring only a small amount, so that this force can be considered nonimpulsive. Momentum for the system of both plates is therefore conserved.

$$(\overset{+}{\leftarrow})\quad m_A(v_A)_1 + m_B(v_B)_1 = m_A(v_A)_2 + m_B(v_B)_2$$

____(1)

$$(\overset{+}{\leftarrow})\quad e = \frac{(v_B)_2 - (v_A)_2}{(v_A)_1 - (v_B)_1}$$

____(2)

Solving Eq. (1) and (2),

$$(v_A)_2 = 0.600 \text{ m/s}$$ **Ans.**

$$(v_B)_2 = 3.40 \text{ m/s}$$ **Ans.**

For plate B:

$$T_2 + V_2 = T_3 + V_3$$

____

$$x = 0.304 \text{ m}$$ **Ans.**

15 - 12. Two disks each have a mass of 2 kg and the initial velocities are shown just before they collide at point O. Determine their speed just after impact. The coefficient of restitution is $e = 0.65$.

Solution

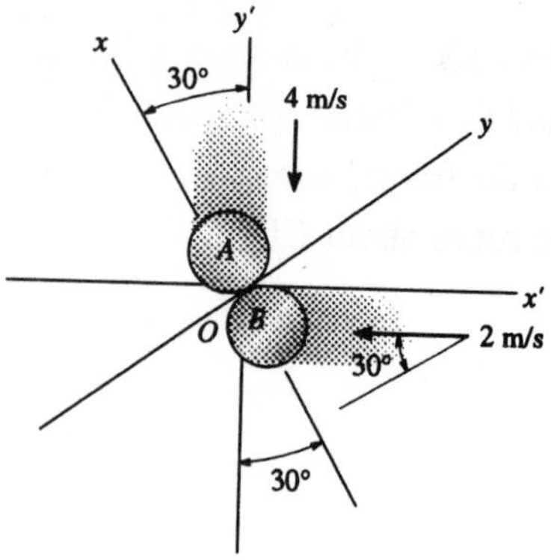

Along the line of impact (x axis)

$$(\nwarrow+)\quad \Sigma m v_1 = \Sigma m v_2$$

$$(\nwarrow+)\quad e = \frac{(v_B)_{2x} - (v_A)_{2x}}{(v_A)_{1x} - (v_B)_{1x}}$$

Solving

$$(v_B)_{2x} = -2.683\text{ m/s},\ (v_A)_{2x} = 0.2188\text{ m/s}$$

Disk A :

$$(\nearrow+)\ \Sigma m(v_A)_{1y} = \Sigma m(v_A)_{2y}$$

$$(v_A)_{2y} = -2\text{ m/s}$$

Disk B :

$$(\nearrow+)\ \Sigma m(v_B)_{1y} = \Sigma m(v_B)_{2y}$$

$$(v_B)_{2y} = -1.732\text{ m/s}$$

$$(v_B)_2 = \sqrt{(-2.683)^2 + (-1.732)^2} = 3.19\text{ m/s}$$ **Ans.**

$$(v_A)_2 = \sqrt{(0.2188)^2 + (-2)^2} = 2.01\text{ m/s}$$ **Ans.**

15 - 13. The two small blocks A and B each have a mass of 0.5 kg. The blocks are fixed to the horizontal rods and their initial speed is 2 m/s. If a couple moment of $M = (0.5t)$ N · m, where t is in seconds, is applied about CD of the frame, determine the speed of the blocks in 2 s. The mass of the supporting frame is negligible and it is free to rotate about CD.

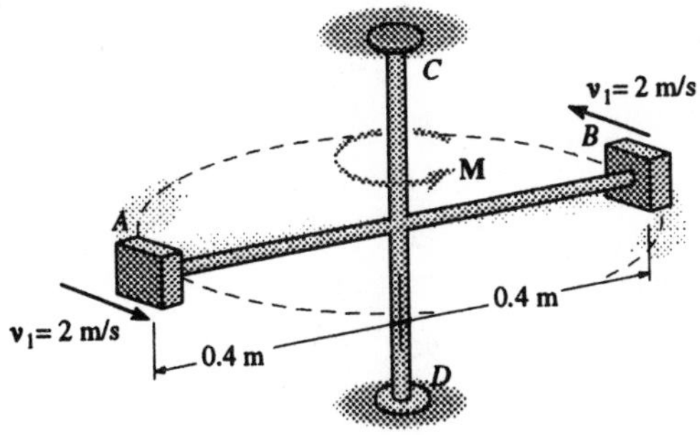

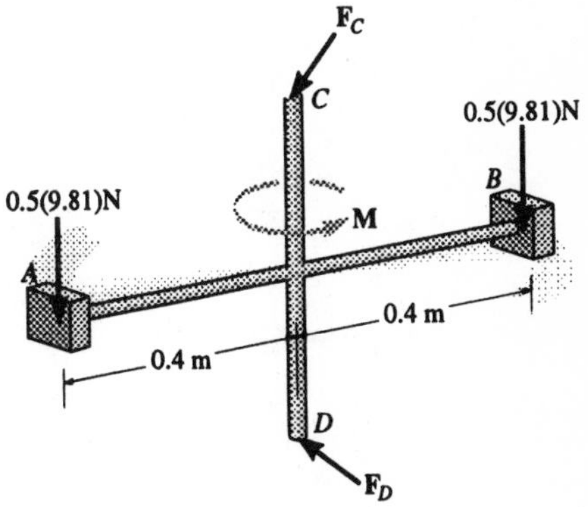

Solution

From the free - body diagram, if we sum moments about the z axis, we can eliminate the reactions F_C , F_D and the weight of the block, and obtain the result.

$$(H_z)_1 + \Sigma \int M_z \, dt = (H_z)_2$$

$v_2 = 4.50$ m/s **Ans.**

15-14. The 50-kg boy holds onto a ring as he runs in a circle and then lifts his feet off the ground as shown. If *initially* his center of gravity G is $r_A = 3$ m from the pole and his velocity is *horizontal* such that $v_A = 2.5$ m/s, determine (a) his velocity when he is at B, where $r_B = 2$ m, $\Delta z = 1$ m, and (b) the vertical component of his velocity, $(\mathbf{v}_B)_z$, which is causing him to fall downward.

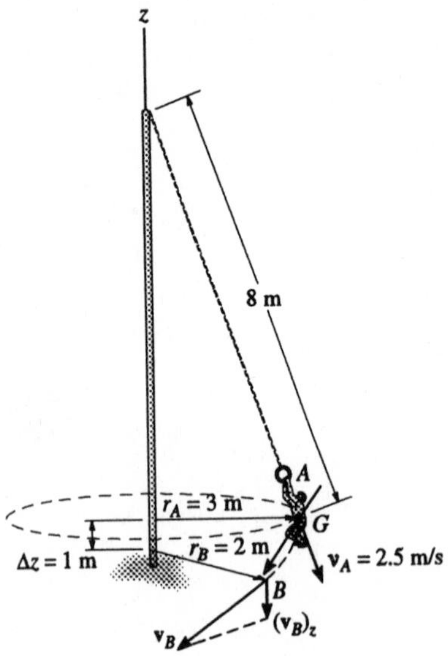

Solution

Put the datum through G when the boy is at A.

$$T_A + V_A = T_B + V_B$$

____

$$v_B = 5.09 \text{ m/s}$$ **Ans.**

$$(H_z)_1 = (H_z)_2$$

____

$$(v_B)_{\text{horiz.}} = 3.75 \text{ m/s}$$

$(v_B)_z =$ ______________________________

$$(v_B)_z = 3.44 \text{ m/s}$$ **Ans.**

16 Planar Kinematics of a Rigid Body

Rotation About a Fixed Axis

16 - 1. The blades of a fan are given an angular acceleration of $\alpha = (0.2\theta)$ rad/s^2, where θ is in radians. If initially the blades have an angular velocity of 5 rad/s, determine the speed of point P located at the tip of one of the blades just after the blade has turned two revolutions.

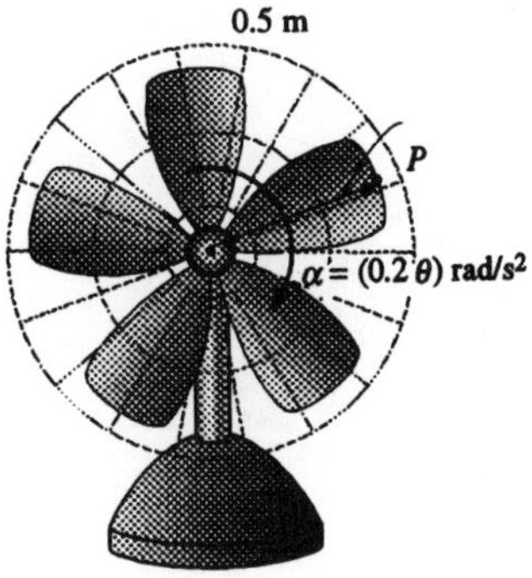

Solution

$$\theta_2 = 2 \text{ rev} = 2\pi(2) = 4\pi \text{ rad}$$

Here α is a function of θ. Set up the definite integrals needed to obtain ω.

____

$$\omega = 7.52 \text{ rad/s}$$

$v_P =$ ____________________

$v_P = 3.76$ m/s **Ans.**

16 - 2. Arm $ABCD$ is pinned at B and undergoes reciprocating motion such that $\theta = (0.3 \sin 4t)$ rad, where t is in seconds and the argument for the sine is in degrees. Determine the largest speed of point A during the motion and the magnitude of the acceleration of point D at this instant.

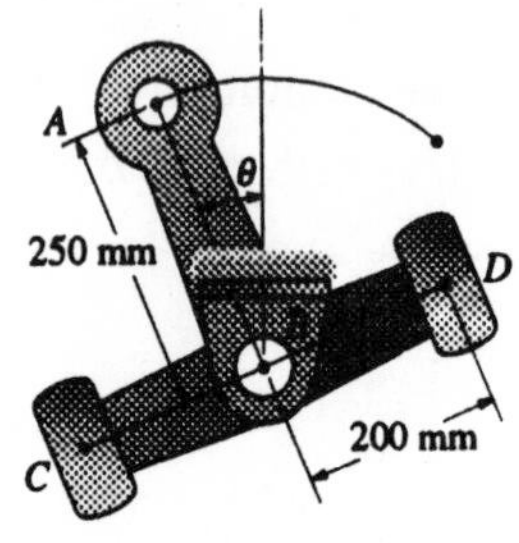

Solution

$\theta = 0.3 \sin 4t$

$\omega =$ ________________________

$\omega_{max} =$ ________________________

$\alpha =$ ________________________

ω_{max} occurs when $\cos 4t = 1$ so that $\sin 4t = 0$. Thus

$\alpha =$ ________________________

$(v_A)_{max} =$ ________________________

$(v_A)_{max} = 300 \text{ mm/s}$ **Ans.**

$(a_D)_t =$ ________________________

$(a_D)_n =$ ________________________

$a_D = 288 \text{ mm/s}^2$ **Ans.**

16 - 3. The wheel starts rotating from rest with an angular acceleration of $\alpha = (3t^2)$ rad/s^2, where t is in seconds. Determine the angular displacement of the gear and the magnitude of the velocity and acceleration of point P when $t = 2$ s.

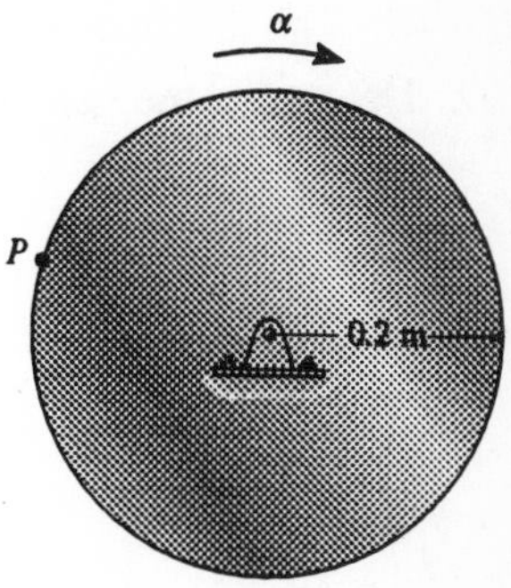

Solution

Here α is a function of t. Set up the definite integrals to find ω as a function of time.

____

$$\omega = (t^3) \text{ rad/s}$$

Use this expression to set up the definite integrals for finding θ as a function of time.

____

$$\theta = (\tfrac{1}{4}t^4) \text{ rad}$$

When $t = 2$ s

$\alpha =$ ____________________

$\omega =$ ____________________

$\theta =$ ____________________ **Ans.**

Now that the angular motions are known, the speed and components of acceleration of P can be found.

$v_P =$ ____________________ $= 1.60$ m/s **Ans.**

$(a_P)_n =$ ____________________ $= 12.8$ m/s^2

$(a_P)_t =$ ____________________ $= 2.40$ m/s^2

Thus,

$a_P =$ ____________________ $= 13.0$ m/s^2 **Ans.**

Absolute General Plane Motion Analysis

16 - 4. The mechanism is used to convert the *constant* circular motion of rod AB into translating motion of rod CD. Determine the velocity and acceleration of CD for any angle θ of AB.

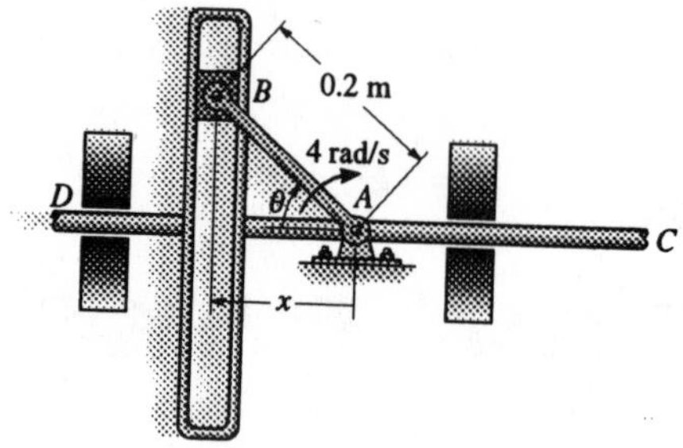

Solution

$x =$ ____________________

$\dot{x} =$ ____________________

$\ddot{x} =$ ____________________

Since

$\dot{\theta} =$ __________

$\ddot{\theta} =$ __________

$v_{CD} = -0.8 \sin\theta$ m/s **Ans.**

$a_{CD} = -3.2 \cos\theta$ m/s^2 **Ans.**

16 - 5. If the angular velocity of AB is maintained at $\omega = 5$ rad/s, determine the required speed v of CD for any angle θ of rod AB.

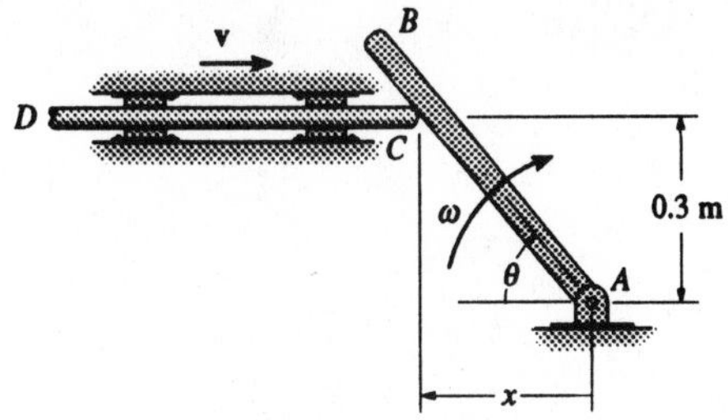

Solution

$x =$ ______________________________

$x =$ ______________________________

$v = -1.5 \csc^2\theta$ **Ans.**

16 - 6. Determine the velocity of the board if the rollers do not slip and have an angular velocity ω.

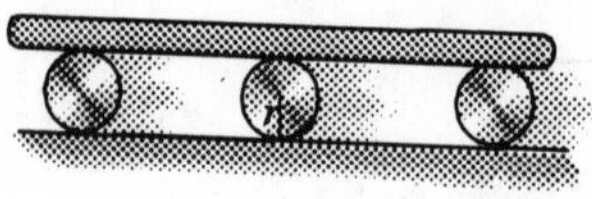

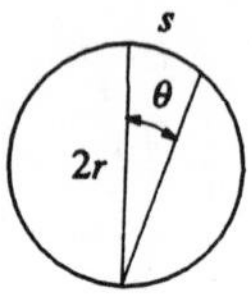

Solution

From the figure

$s =$ ______________________________

$v =$ ______________________________ **Ans.**

16 - 7. If A is approaching B with a speed of 0.5 m/s, determine the speed at which the platform is rising as a function of θ. Each link is pin - connected at its midpoint and end points and has a length of 1.5 m.

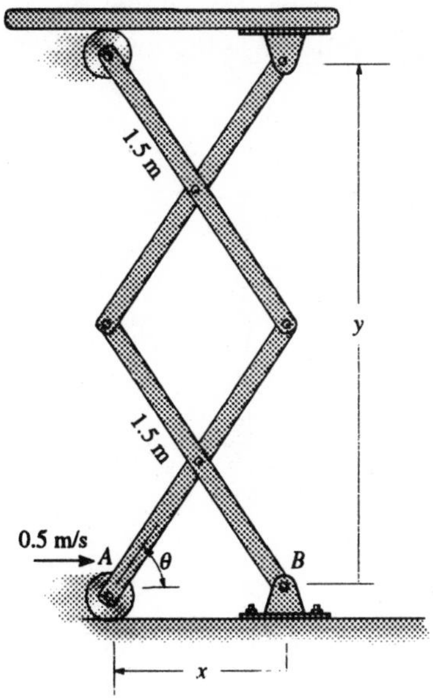

Solution

Relate y to θ,

$y =$ ____________________________________

$\dot{y} =$ ____________________________________ (1)

Relate x to θ,

$x =$ ____________________________________

$\dot{x} =$ ____________________________________ (2)

Eliminating $\dot{\theta}$ between Eqs. (1) and (2) yields

$$\dot{y} = -\dot{x} \cot \theta$$

$$v_S = -v_A \cot \theta$$

$$v_S = 0.5 \cot \theta \text{ m/s}$$ **Ans.**

16-8. The bar is confined to move along the horizontal and vertical planes. If the velocity of its end A is 2 m/s, determine the bar's angular velocity and the velocity of point B at the instant $\theta = 60°$.

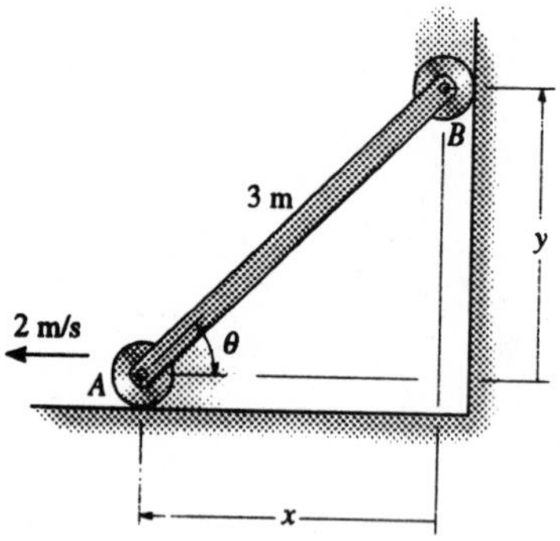

Solution

Relate x to θ,

$x =$ ________________________________

$v_A =$ ________________________________

When $v_A = 2$ m/s, $\theta = 60°$,

$\omega = -0.7968$ rad/s $= 0.7968$ rad/s **Ans.**

Relate y to θ,

$y =$ ________________________________

$v_B =$ ________________________________

Substituting $\omega = \dot{\theta} = -0.7968$ rad/s, $\theta = 60°$,

$v_B = 1.15$ m/s $\downarrow$ **Ans.**

16 - 9. Due to an engine failure, the missile is rotating at $\omega = 3$ rad/s, while its mass center G is moving upward at 200 m/s. Determine the velocity of its nose B at this instant.

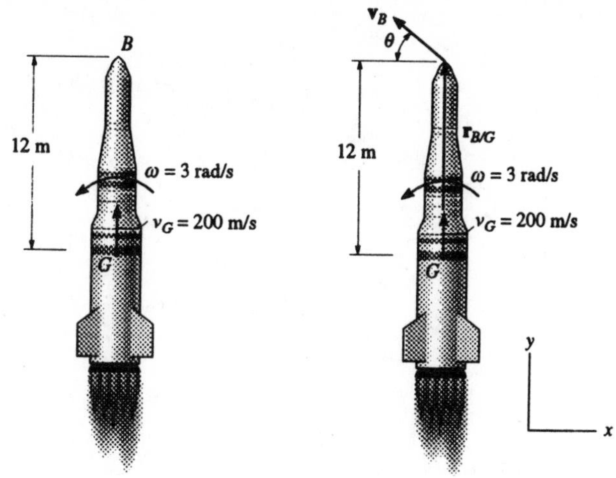

Solution

Apply the relative velocity equation using Cartesian vectors.

$$\mathbf{v}_B = \mathbf{v}_G + \omega \times \mathbf{r}_{B/G}$$

$(v_B)_x\mathbf{i} + (v_B)_y\mathbf{j} =$ ______________________________

$(v_B)_x = -3(12) = -36 \text{ m/s} = 36 \text{ m/s} \leftarrow$

$(v_B)_y = 200 \text{ m/s} \uparrow$

$v_B =$ ______________________________

$v_B = 203$ m/s **Ans.**

$\theta = \tan^{-1}\left(\frac{200}{36}\right) = 79.8°$ **Ans.**

16 - 10. If the block at C is moving downward at 4 m/s, determine the angular velocity of bar AB at the instant shown.

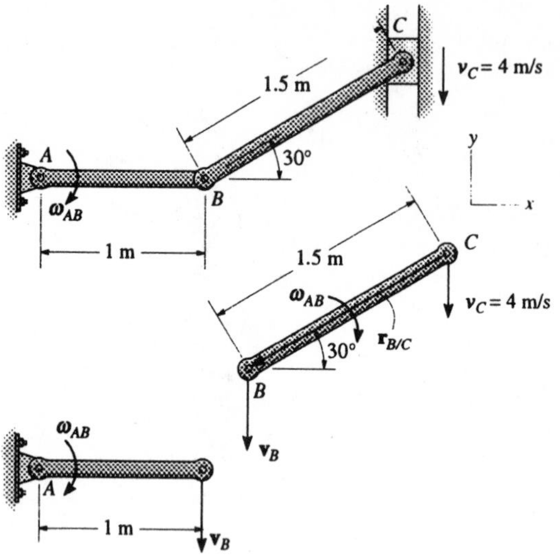

Solution

Since AB is subjected to rotation about a fixed axis the velocity of B is

$v_B =$ ____________________ $\downarrow$

Apply the relative velocity equation to bar BC using Cartesian vectors.

$$\mathbf{v}_B = \mathbf{v}_C + \omega_{BC} \times \mathbf{r}_{B/C}$$

____

$$0 = -1.5\omega_{BC}\sin 30°$$

$$-\omega_{AB}(1) = -4 + 1.5\omega_{BC}\cos 30°$$

Solving,

$$\omega = 0$$

$$\omega_{AB} = 4 \text{ rad/s}$$ **Ans.**

16 - 11. Knowing the angular velocity of link CD is $\omega_{CD} = 4$ rad/s, determine the angular velocity of links BC and AB at the instant shown.

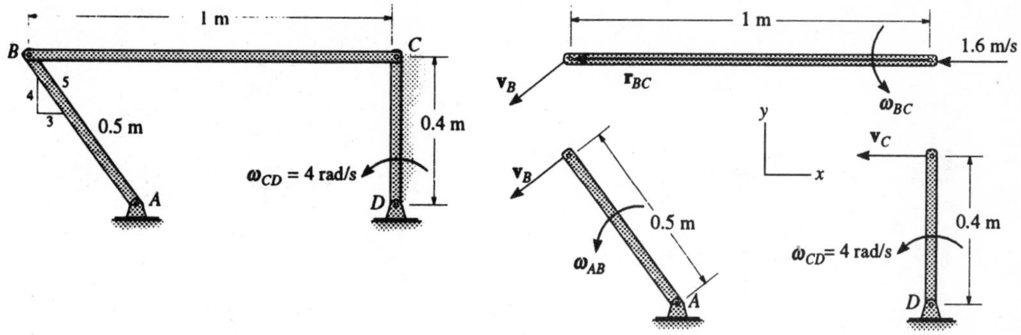

Solution

Since CD is subjected to rotation about a fixed axis, the velocity of C is

$$v_C = \_\_\_\_\_\_\_\_\_\_\_\_\_\_\_\_\_\_\_\_\_\_\_\_\_\_\_ = 1.6 \text{ m/s} \leftarrow$$

Apply the relation velocity equation to bar BC using Cartesian vectors.

$$\mathbf{v}_B = \mathbf{v}_C + \omega_{BC} \times \mathbf{r}_{B/C}$$

____

$$-\left(\frac{4}{5}\right)v_B = -1.6$$

$$-\left(\frac{3}{5}\right)v_B - 1\omega_{BC}$$

Solving

$v_B = 2$ m/s **Ans.**

$\omega_{BC} = 1.20$ rad/s **Ans.**

Knowing v_B, determine ω_{AB}

$\omega_{AB} = \_\_\_\_\_\_\_\_\_\_\_\_\_\_\_\_\_\_\_\_\_\_\_\_\_\_\_$ rad/s **Ans.**

16 - 12. If rod *CD* has a downward velocity of 6 m/s at the instant shown, determine the velocity of the gear rack *G* at this instant. The rod is pinned at *C* to gear *B*.

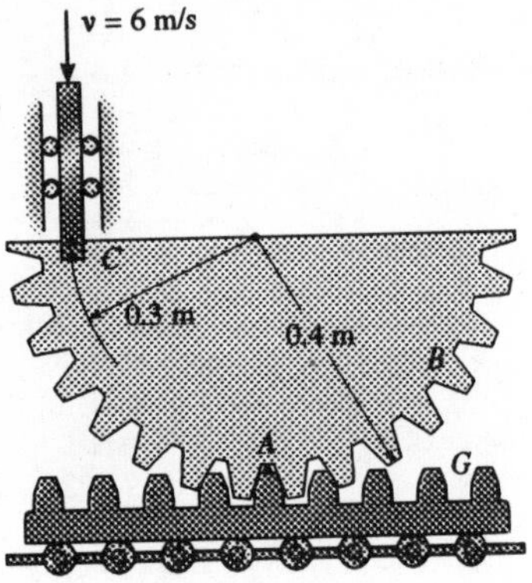

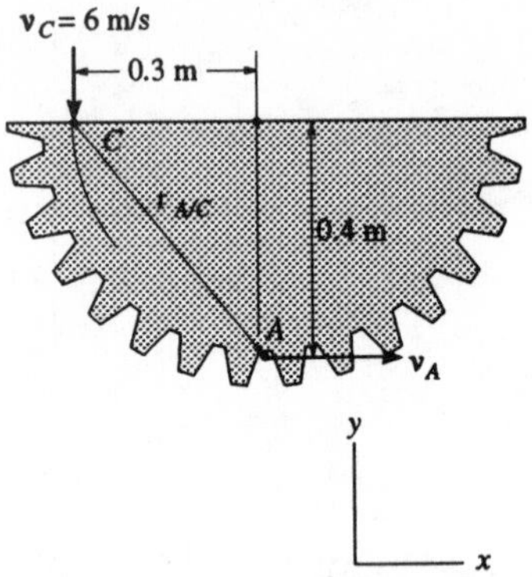

Solution

Apply the relative velocity eqation between points *A* and *C*. Use Cartesian vectors.

$$\mathbf{v}_A = \mathbf{v}_C + \omega \times \mathbf{r}_{A/C}$$

$$v_A = 0.4\omega$$

$$0 = -6 + 0.3\omega$$

Solving,

$$\omega = 20 \text{ rad/s}$$

$$v_A = 8 \text{ m/s}$$ **Ans.**

16-13. If AB has an angular velocity of $\omega_{AB} = 8$ rad/s, determine the angular velocity of gear F at the instant shown. Gear E is a part of arm CD.

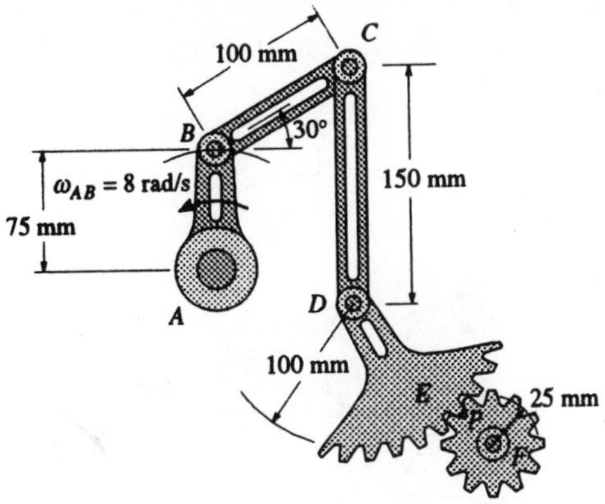

Solution

$v_B =$ ______________________ $= 600$ mm/s

Apply the relative velocity equation between points B and C. Use Cartesian vectors.

$$\mathbf{v}_C = \mathbf{v}_B + \boldsymbol{\omega}_{BC} \times \mathbf{r}_{C/B}$$

____

$$-v_C = -600 - \omega_{BC}(100 \sin 30°)$$

$$0 = \omega_{BC}(100 \cos 30°)$$

Solving,

$$\omega_{BC} = 0$$

$$v_C = 600 \text{ mm/s}$$

Since E rotates about point D, determine the angular velocity of E.

$\omega_E =$ ______________________________

The speed of a point P in contact between gears E and F is

$v_P =$ ______________________________

The angular velocity of gear F is therefore

$\omega_F =$ ______________________ $= 16$ rad /s **Ans.**

Instantaneous Center of Zero Velocity

16-14. The automobile is traveling in a straight path at 12 m/s. If no slipping occurs, determine the angular velocity of one of the rear wheels and the velocity of the fastest moving point on the wheel.

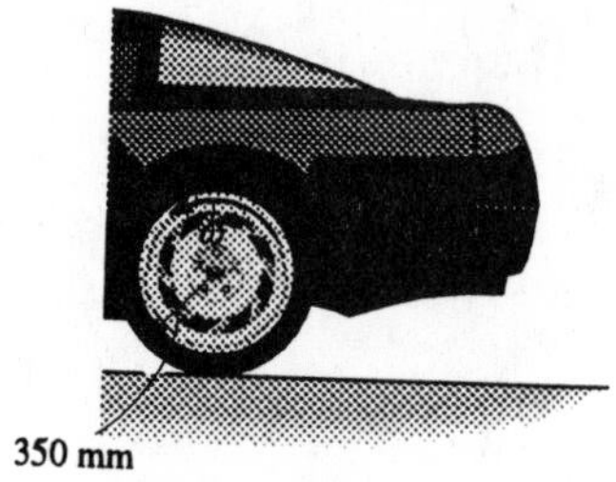

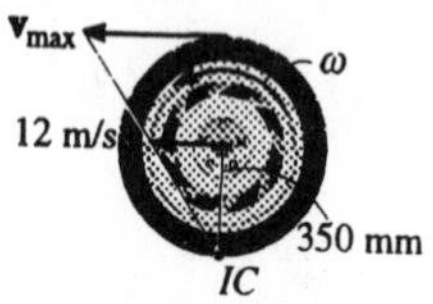

Solution

The *IC* is at the ground since no slipping occurs there.

$\omega =$ ______________________________

$\omega = 34.3$ rad/s **Ans.**

$v_{max} =$ ______________________________

$v_{max} = 24$ m/s **Ans.**

16 - 15. Part of an automatic transmission consists of a *fixed* ring gear R, three equal planet gears P, the sun gear S, and the planet carrier C, which is shaded. If the sun gear is rotating at $\omega_S = 4$ rad/s, determine the angular velocity of the *planet carrier*. Note that C is pin - connected to the center of each of the planet gears.

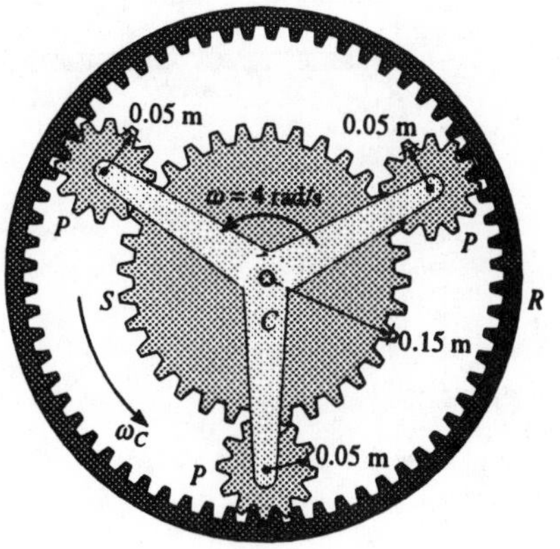

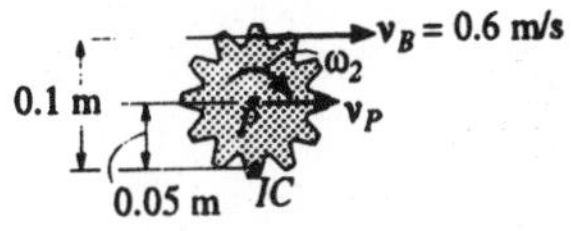

Solution

The IC for each planet gear is at the contact point with the ring gear. The velocity of the contact point B with the sun gear can be determined.

$v_B =$ ____ $= 0.6$ m/s

The angular velocity of the planet gear is

$\omega_P =$ ____ $= 6$ rad/s

Thus

$v_P =$ ____ $= 0.3$ m/s

The angular velocity of the planet carrier is,

$\omega_C =$ ____ $= 1.50$ rad/s **Ans.**

16 - 16. If the crank DC of the mechanism rotates at a constant rate of 6 rad/s, determine the speed of point A.

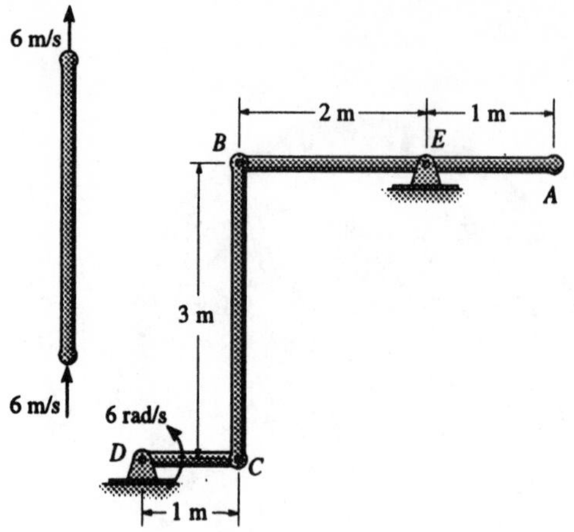

Solution

By inspection, DC and AB are subjected to motion about a fixed axis. Calculate

$v_C =$ ______________________________

Note that point B follows a circular path about the pin E and therefore the velocity of B is also vertical. The IC is at infinity. Thus,

$\omega_{BC} =$ __________

$v_B =$ __________

$\omega_{BA} =$ ____________________ $= 3$ rad/s

$v_A = 3$ m/s **Ans.**

16 - 17. The pulley is pin - connected to block *B* at *A*. As cord *CF* unwinds from the inner hub with the motion shown, cord *DE* unwinds from the outer rim. Determine the angular acceleration of the pulley at the instant shown.

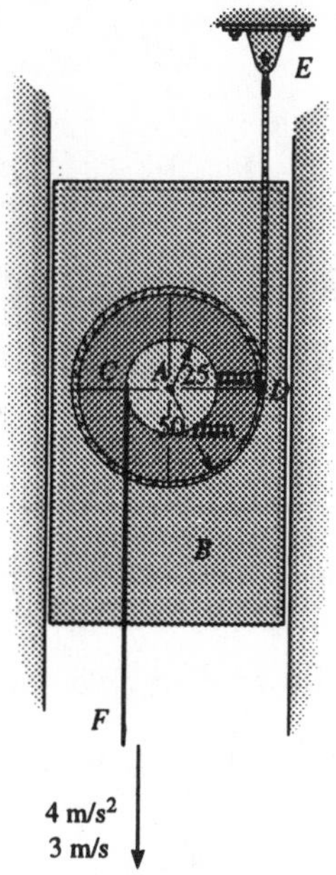

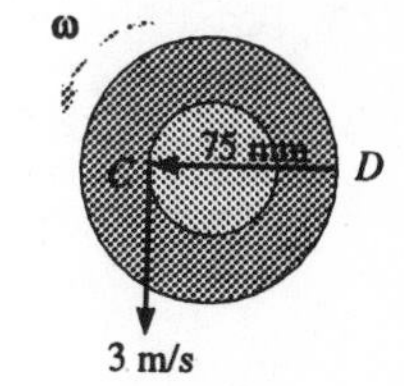

Solution

Velocity Analysis. We can determine the angular velocity of the pulley from the *IC*.

$\omega =$ ____ $= 40$ rad/s

Acceleration Analysis. Apply the relative acceleration equation to the pulley using Cartesian vectors. Note that both points *C* and *D* have normal components of acceleration since these points move along curved pathes.

$$\mathbf{a}_C = \mathbf{a}_D + \alpha \times \mathbf{r}_{C/D} - \omega^2 \mathbf{r}_{C/D}$$

____

$(a_C)_x = (40)^2(0.075)$

$-4 = -\alpha(0.075)$

$\alpha = 53.3 \text{ rad/s}^2$ **Ans.**

16 - 18. The disk rolls without slipping such that it has an angular acceleration of $\alpha = 4\ \text{rad/s}^2$ and angular velocity of $\omega = 2\ \text{rad/s}$ at the instant shown. Determine the accelerations of points A and B on the link and the link's angular acceleration at this instant. Assume point A lies on the periphery of the disk, 150 mm from C.

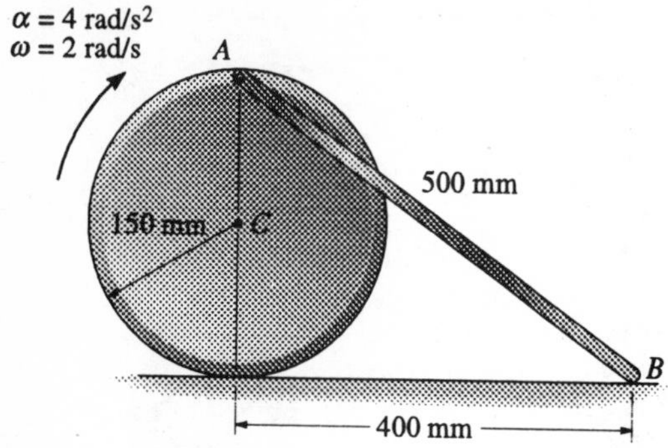

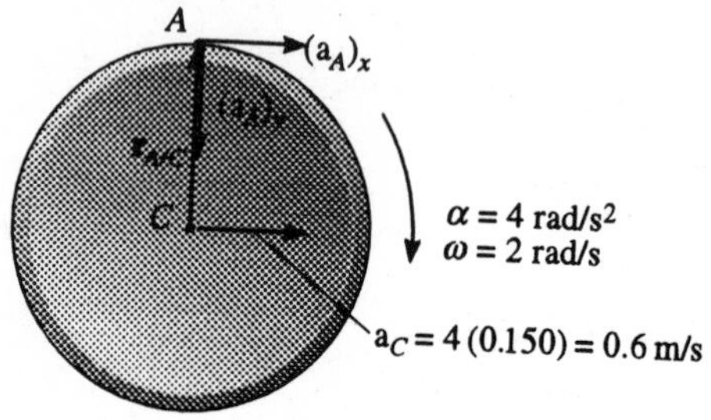

Solution

Velocity Analysis. At the instant shown the link is translating and so A and B move horizontally. Thus the IC is at infinity and so $\omega_{AB} = 0$.

Acceleration Analysis. We can determine the acceleration of point A by applying the relative acceleration equation between points A and C on the disk.

$$\mathbf{a}_A = \mathbf{a}_C + \alpha \times \mathbf{r}_{A/C} - \omega^2 \mathbf{r}_{A/C}$$

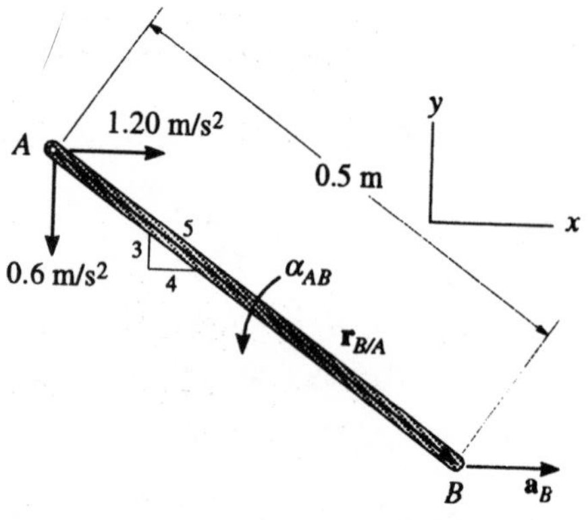

$$(a_A)_x = 0.6 + 4(0.15)$$

$$(a_A)_y = -(2)^2(0.15)$$

$(a_A)_x = 1.20\ \text{m/s}^2 \rightarrow$ **Ans.**

$(a_A)_y = 0.6\ \text{m/s}^2 \downarrow$ **Ans.**

Using these results one can now apply the relative acceleration equation between points A and B on the link.

$$\mathbf{a}_B = \mathbf{a}_A + \alpha_{AB} \times \mathbf{r}_{B/A} - \omega_{AB}^2 \mathbf{r}_{B/A}$$

$$a_B = 1.20 + 0.3\alpha_{AB}$$

$$0 = -0.6 + 0.4\alpha_{AB}$$

$\alpha_{AB} = 1.50\ \text{rad/s}^2$ **Ans.**

$a_B = 1.65\ \text{m/s}^2$ **Ans.**

16-19. At the instant shown, arm AB has an angular velocity of $\omega_{AB} = 0.5$ rad/s and an angular acceleration of $\alpha_{AB} = 2$ rad/s^2. Determine the angular velocity and angular acceleration of the link CD at this instant.

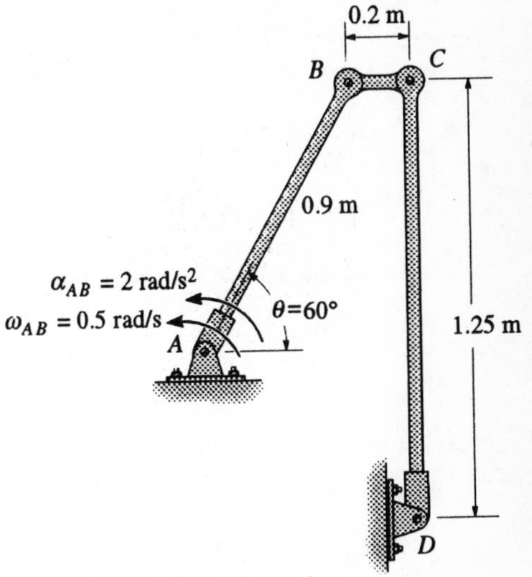

Solution

Velocity Analysis. From rod AB,

$v_B =$ ______________________ $= 0.45$ m/s

The direction of the velocity of points B and C are known and so the location of the IC can be determined.

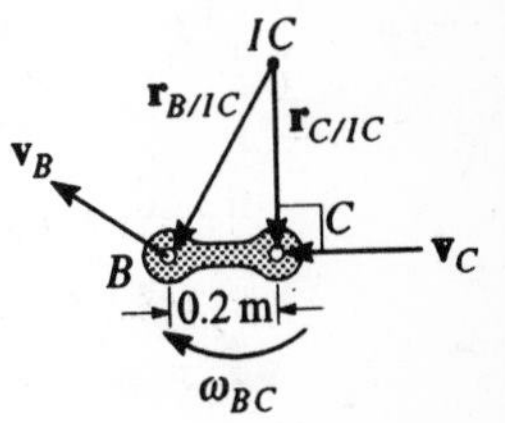

$\omega_{BC} =$ ______________________ $= 1.125$ rad/s

$v_C =$ ______________________ $= 0.390$ m/s $\leftarrow$

We can now determine the angular velocity of link CD.

$\omega_{CD} =$ ______________________ $= 0.312$ rad/s **Ans.**

Acceleration Analysis. From rod AB,

$(a_B)_t =$ ______________________ $= 1.80$ m/s^2

$(a_B)_n =$ ______________________ $= 0.225$ m/s^2

Since ω_{DC} is known, the normal component of the acceleration of point C can be determined.

$(a_C)_n =$ ______________________ $= 0.122$ m/s^2

Applying the relative acceleration equation between points B and C.

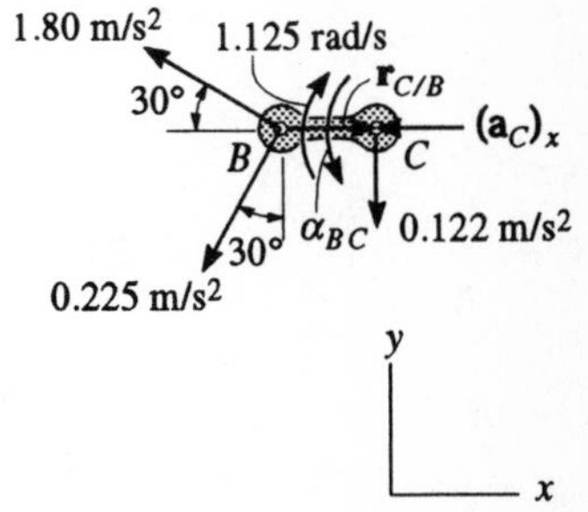

$$\mathbf{a}_C = \mathbf{a}_B + \alpha_{BC} \times \mathbf{r}_{C/B} - (\omega_{BC})^2 \mathbf{r}_{C/B}$$

____

$$-(a_C)_x = -1.80 \cos 30° - 0.225 \sin 30° - (1.125)^2(0.2)$$
$$-0.122 = 1.80 \sin 30° - 0.225 \cos 30° + \alpha_{BC}(0.2)$$

Solving,

$$(a_C)_x = 1.92 \text{ m/s}^2; \quad \alpha_{BC} = -4.14 \text{ rad/s}^2$$

Thus, the angular acceleration of link CD is

$\alpha_{DC} =$ ______________________ $= 1.54$ rad/s^2 **Ans.**

17 Planar Kinetics of a Rigid Body: Force and Acceleration

Equations of Motion: Translation

17 - 1. The 10 - kg bar is fixed to the carriage at A. Determine the internal axial force N_y, shear force V_x, and moment M_A at A if the carriage is descending the plane with an acceleration of $4\ \text{m/s}^2$.

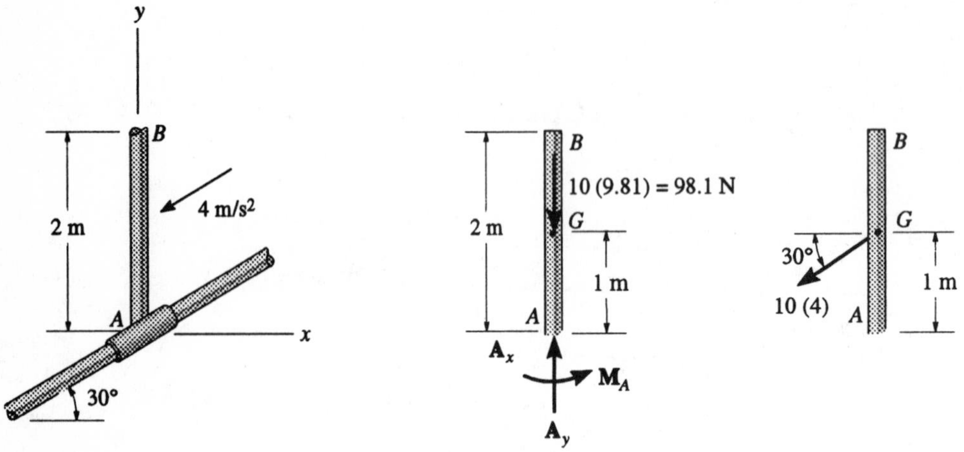

Solution

The free - body and kinetic diagrams of the bar are drawn first.

$+\uparrow \Sigma F_y = m(a_G)_y;$ ____________________

$N_y = 78.1\ \text{N}$ **Ans.**

$\overset{+}{\leftarrow} \Sigma F_x = m(a_G)_x;$ ____________________

$V_x = 34.6\ \text{N}$ **Ans.**

$\circlearrowleft + \Sigma M_A = \Sigma (M_k)_A;$ ____________________

$M_A = 34.6\ \text{N} \cdot \text{m}$ **Ans.**

17 - 2. The dragster has a mass of 1.3 Mg and a center of mass at G. A parachute is attached at C and provides a horizontal braking force of $F = (1.8v^2)$ N, where v is in m/s. Determine the deceleration the dragster can have upon releasing the parachute such that the wheels at B are on the verge of leaving the ground i.e., the normal reaction at B is zero. Neglect the mass of the wheels and assume the engine is disengaged so that the wheels are free to roll.

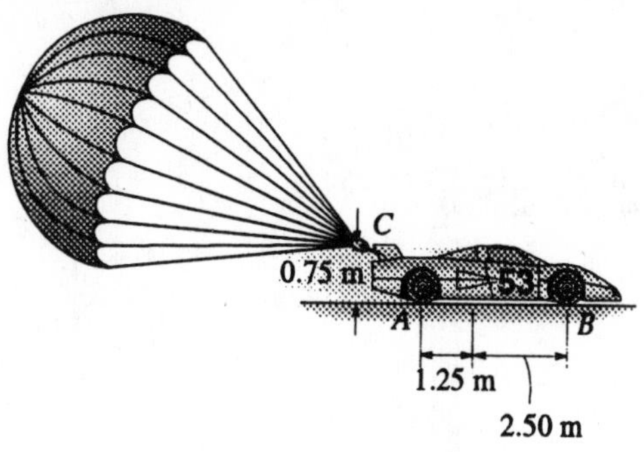

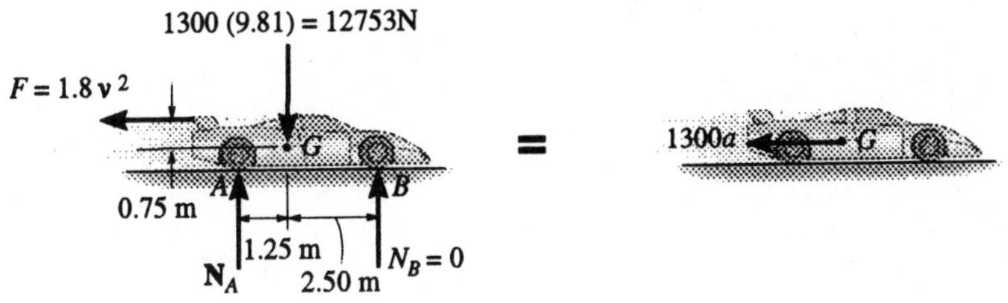

Solution

The free - body diagram is drawn first.

$\overset{+}{\leftarrow} \Sigma F_x = m(a_G)_x;$ ____

$\circlearrowleft + \Sigma M_G = 0;$ ____

$+\uparrow \Sigma F_y = m(a_G)_y;$ ____

Solving,

$N_A = 12.7$ kN

$v = 109$ m/s

$a_G = 16.4$ m/s^2 **Ans.**

17 - 3. The 10 - kg block rests on the platform for which the coefficient of static friction is $\mu_s = 0.4$. If at the instant shown link AB has an angular velocity $\omega = 2$ rad/s, determine the greatest angular acceleration of the link so that the block doesn't slip.

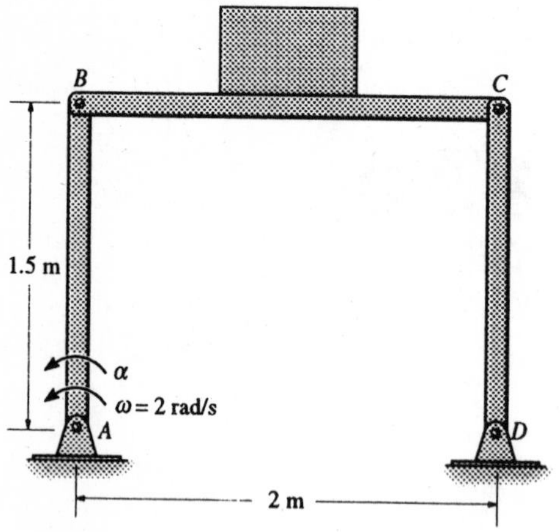

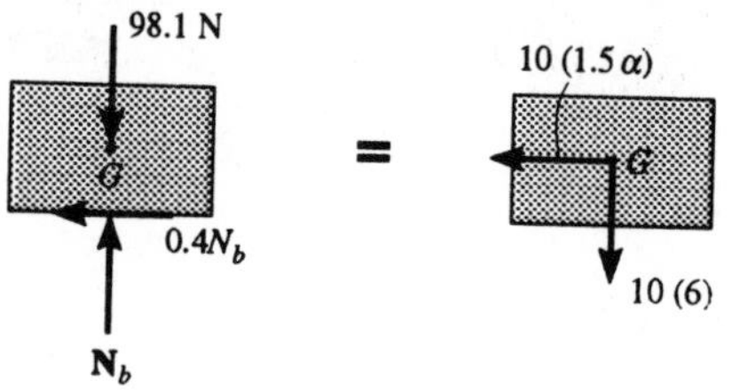

Solution

The block is subjected to curvilinear translation. The acceleration of its mass center can be calculated.

$$(a_G)_t = (a_B)_t = (\alpha)(1.5)$$

$$(a_G)_n = (a_B)_n = (2)^2(1.5) = 6 \text{ m/s}^2$$

The free - body and kinetic diagrams are drawn first.

$+\uparrow \Sigma F_n = m(a_G)_n;$ ________________________________

$$N_b = 38.1 \text{ N}$$

$\overset{+}{\leftarrow} \Sigma F_t = m(a_G)_t;$ ________________________________

$$\alpha = 1.02 \text{ rad/s}^2$$ **Ans.**

17 - 4. The 100 - kg crate C rests on the floor of an elevator for which the coefficient of static friction is $\mu_s = 0.4$. Determine the largest initial angular acceleration α, starting from rest, which the parallel links AB and DE can have without causing the crate to slip. No tipping occurs.

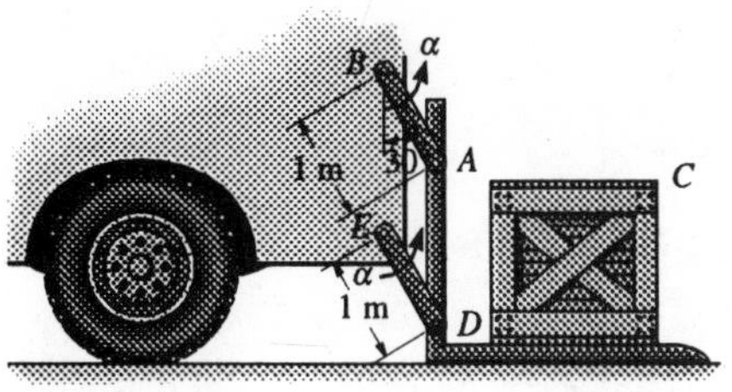

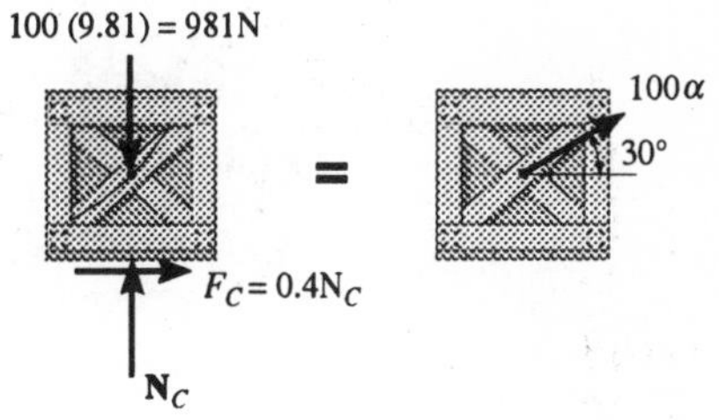

Solution

The crate undergoes curvilinear acceleration. The acceleration of its mass center can be calculated.

$(a_G)_n = (a_C)_n =$ ____________________

$(a_G)_t = (a_C)_t =$ ____________________

The free - body and kinetic diagrams are drawn first.

$\overset{+}{\rightarrow} \Sigma F_x = ma_x$; ____________________

$+\uparrow \Sigma F_y = ma_y$; ____________________

Solving

$N_C = 1.28$ kN

$\alpha = 5.89 \text{ m/s}^2$ **Ans.**

17 - 5. The trailer portion of a truck has a mass of 4 Mg with a center of mass at G. If a *uniform* crate, having a mass of 800 kg and a center of mass at G_C, rests on the trailer, determine the horizontal and vertical components of reaction at the ball and socket joint (pin) A when the truck is decelerating at a constant rate of $a = 3\ \text{m/s}^2$. Assume that the crate does not slip on the trailer and neglect the mass of the wheels. The wheels at B roll freely.

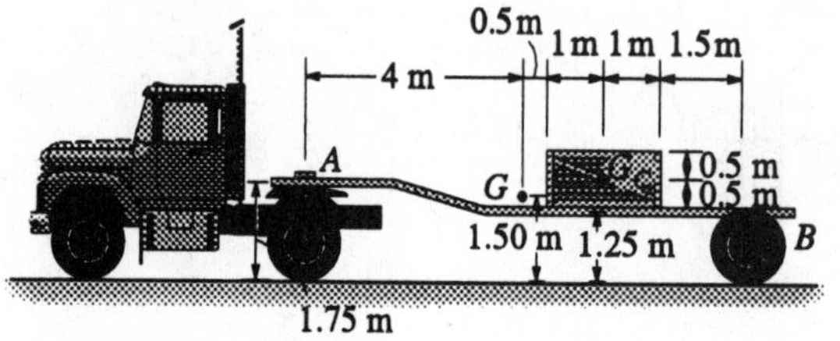

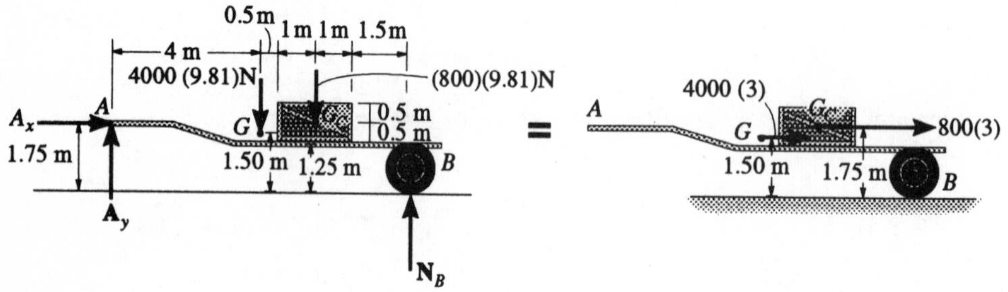

Solution

The free - body and kinetic diagrams are drawn first.

$\overset{+}{\rightarrow} \Sigma F_x = m(a_G)_x;$ ____________________

$A_x = 14\,400\ \text{N} = 14.4\ \text{kN}$ **Ans.**

$\circlearrowleft + \Sigma M_B = \Sigma (M_k)_B;$ ____________________

$A_y = 21\,697.5\ \text{N} = 21.7\ \text{kN}$ **Ans.**

17 - 6. The 15 - kg rod is pinned at its end and has an angular velocity of $\omega = 5$ rad/s when it is in the horizontal position shown. Determine the rod's angular acceleration and the pin reactions at this instant.

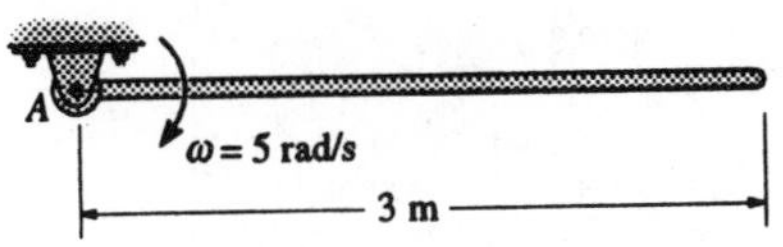

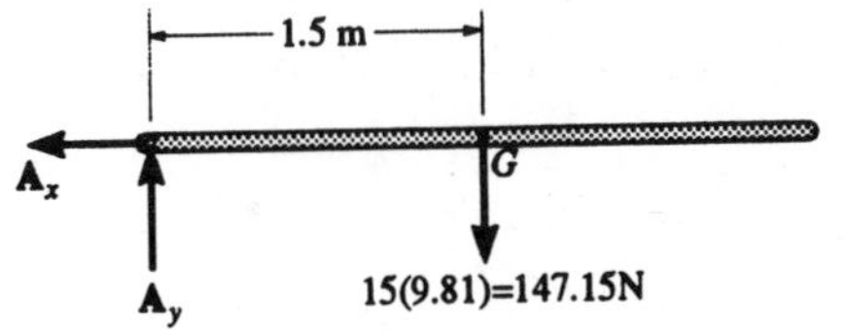

Solution

The free - body diagram is drwn first. Moments will be summed about point A to eliminate A_x and A_y.

$\circlearrowleft + \Sigma M_A = I_A\,\alpha;$ ____

$$\alpha = -4.905 \text{ rad/s}^2 = 4.905 \text{ rad/s}^2 \qquad \textbf{Ans.}$$

$\overset{+}{\leftarrow} \Sigma F_x = m(a_G)_x;$ ____

$$A_x = 562.5 \text{ N} \qquad \textbf{Ans.}$$

Using the result for α, we have

$+\uparrow \Sigma F_y = m(a_G)_y;$ ____

$$A_y = 36.8 \text{ N} \qquad \textbf{Ans.}$$

17 - 7. If the support at B is suddenly removed, determine the initial reactions at the pin A. The plate has a mass of 30 kg.

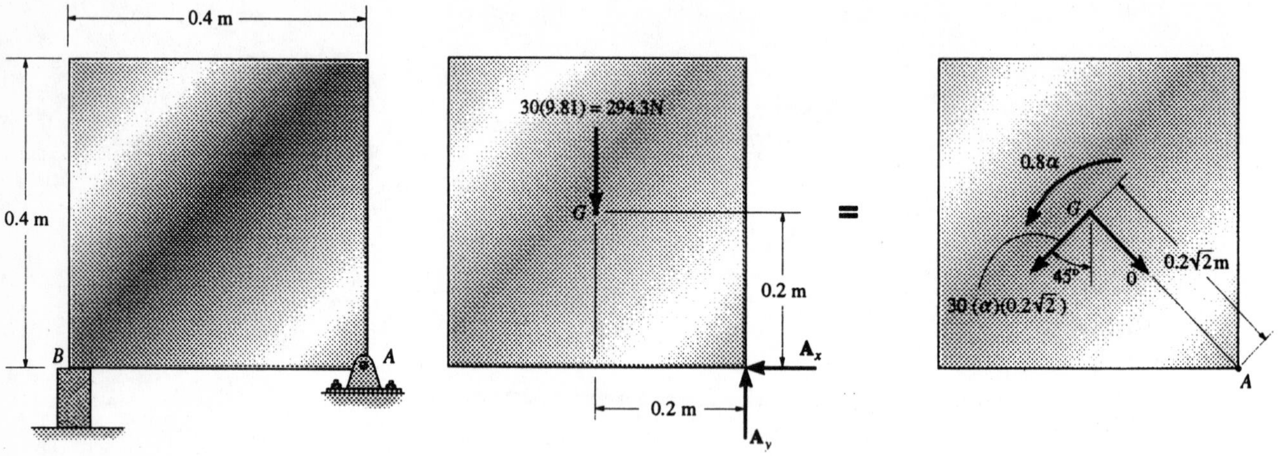

Solution

Initially $\omega = 0$. The free - body and kinetic diagrams are drawn first.

Here $I_G = \frac{1}{12}(30)[(0.4)^2 + (0.4)^2] = 0.8 \text{ kg} \cdot \text{m}^2$

$\overset{+}{\leftarrow} \Sigma F_x = m(a_G)_x;$ ________________________________

$+\uparrow \Sigma F_y = m(a_G)_y;$ ________________________________

$\circlearrowleft + \Sigma M_A = \Sigma(M_k)_A;$ ________________________________

Solving,

$\alpha = 18.4 \text{ rad/s}^2$

$A_x = 110 \text{ N}$ **Ans.**

$A_y = 184 \text{ N}$ **Ans.**

17 - 8. The 80 - kg bar has a mass center at G, and a radius of gyration about G of $k_G = 1.2$ m. Determine the coefficient of static friction at A if it is observed that slipping occurs when $\theta = 30°$. It is released from rest when $\theta = 0°$.

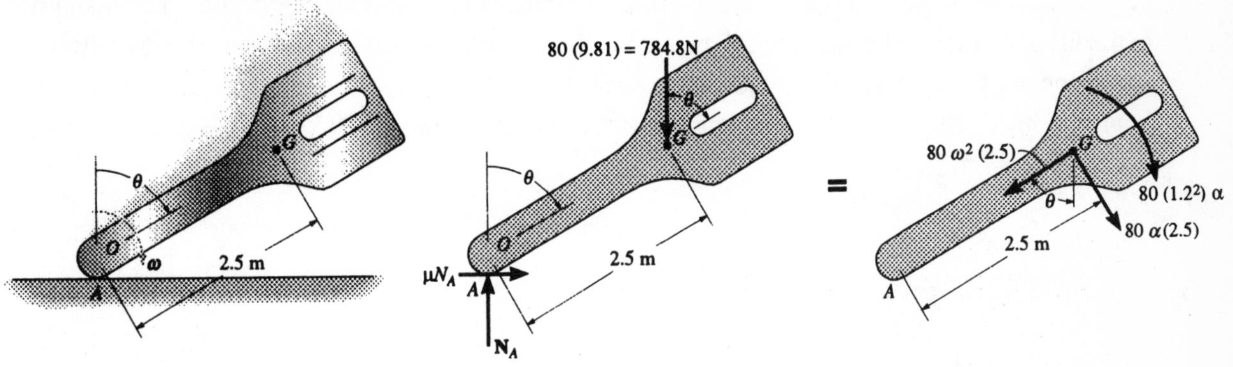

Solution

The free - body and kinetic diagrams are drawn first.

$\overset{+}{\rightarrow} \Sigma F_x = m(a_G)_x$; ____(1)

$+\uparrow \Sigma F_y = m(a_G)_y$; ____(2)

$\curvearrowleft + \Sigma M_A = \Sigma (M_k)_A$; ____

Thus,

$$\alpha = 3.19 \sin\theta$$

Since α is a function of θ, we can use kinematics to relate ω to θ. Set up the definite integrals to do this.

____

$$3.19[-\cos\theta]_0^\theta = \frac{1}{2}\omega^2$$

$$6.78[1 - \cos\theta] = \omega^2$$

At $\theta = 30°$, $\alpha = 1.595 \text{ rad/s}^2$, $\omega = 0.944$ rad/s

From Eqs. (1) and (2),

$N_A = 477$ N **Ans.**

$\mu_A = 0.400$

Equations of Motion: General Plane Motion

17 - 9. A woman sits in a rigid position on her rocking chair by keeping her feet on the bottom rung at B. At the instant shown, she has reached an extreme backward position and has zero angular velocity. Determine her forward angular acceleration α and the frictional force at A necessary to prevent the rocker from slipping. The woman and the rocker have a combined mass of 90 kg and a radius of gyration about G of $k_G = 0.45$ m.

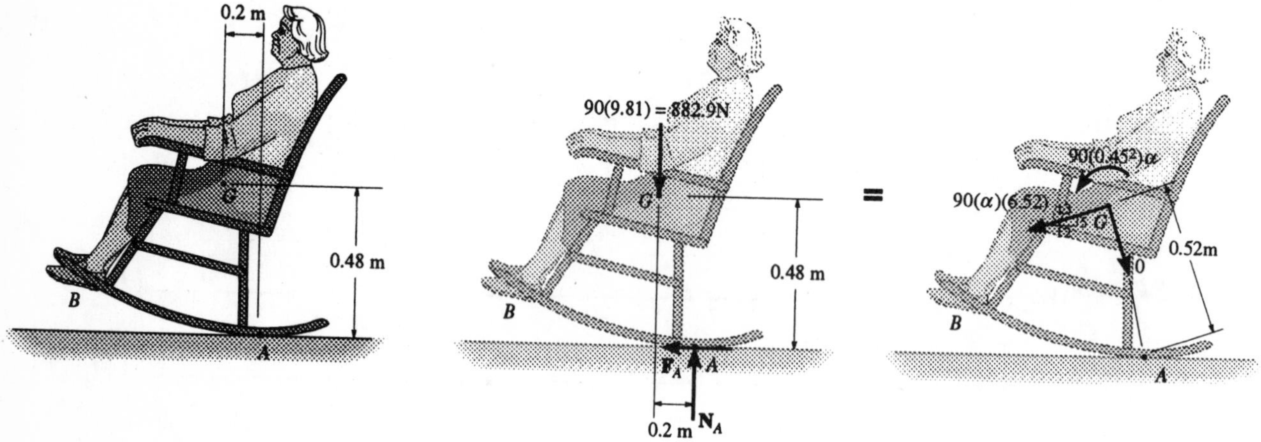

Solution

The free - body and kinetic diagrams are drawn first.

$\overset{+}{\leftarrow} \Sigma F_x = m(a_G)_x$;____

$+\uparrow \Sigma F_y = m(a_G)_y$;____

$\circlearrowleft + \Sigma M_A = \Sigma(M_k)_A$;____

Solving;

$\alpha = 4.15 \text{ rad/s}^2$ **Ans.**

$F_A = 179 \text{ N}$ **Ans.**

$N_A = 808 \text{ N}$

17-10. The slender 200-kg beam is suspended by a cable at its end as shown. If a man pushes on its other end with a horizintal force of 30 N, determine the initial acceleration of its mass center G, the beam's angular acceleration, and the tension in the cable AB.

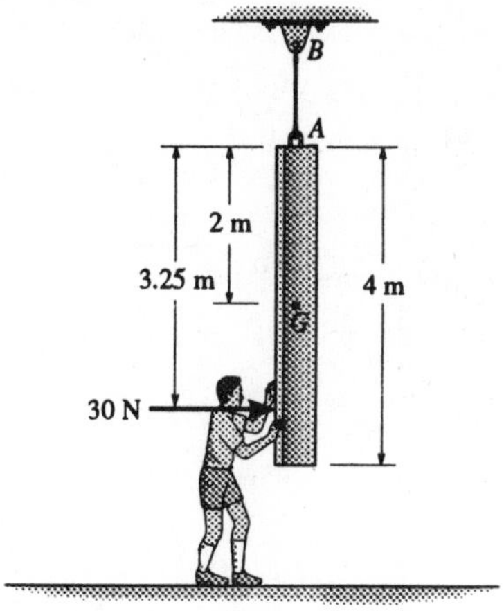

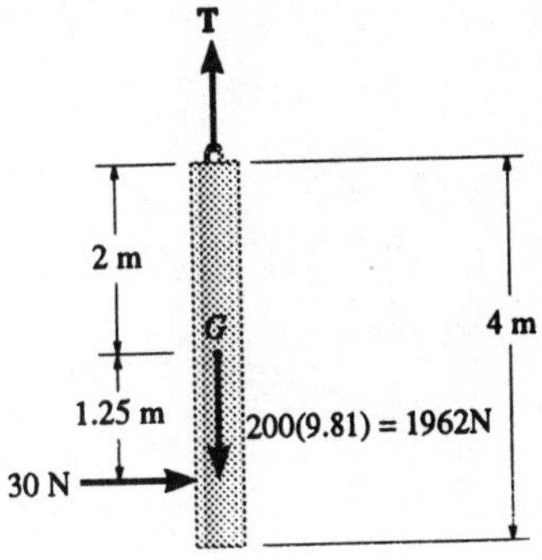

Solution

The free-body diagram of the beam is drawn first.

$\overset{+}{\rightarrow} \Sigma F_x = m(a_G)_x$;________________________________

$+\uparrow \Sigma F_y = m(a_G)_y$;________________________________

$\circlearrowleft + \Sigma M_G = I_G \alpha$;________________________________

Solving;

$T = 1962$ N **Ans.**

$a_G = 0.150 \text{ m/s}^2$ **Ans.**

$\alpha = 0.141 \text{ rad/s}^2$ **Ans.**

17 - 11. A spool and the telephone wire wrapped around its core have a total mass of 80 kg and a radius of gyration of $k_G = 0.3$ m. If the coefficient of static friction between the spool and the ground is $\mu = 0.4$, and the coefficient of kinetic friction is $\mu_k = 0.2$, determine the angular acceleration of the spool if the end of the cable is subjected to a horizontal force of 30 N.

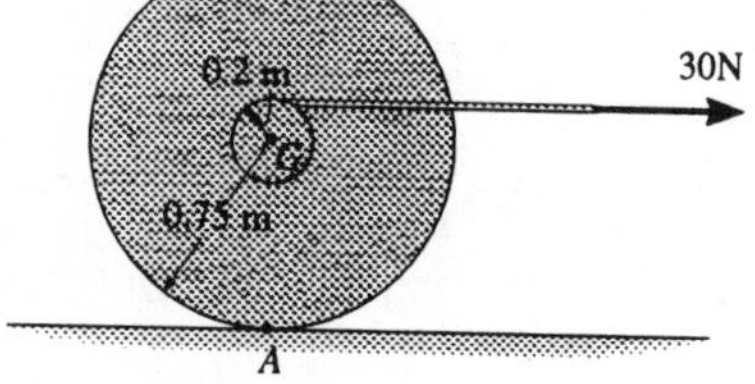

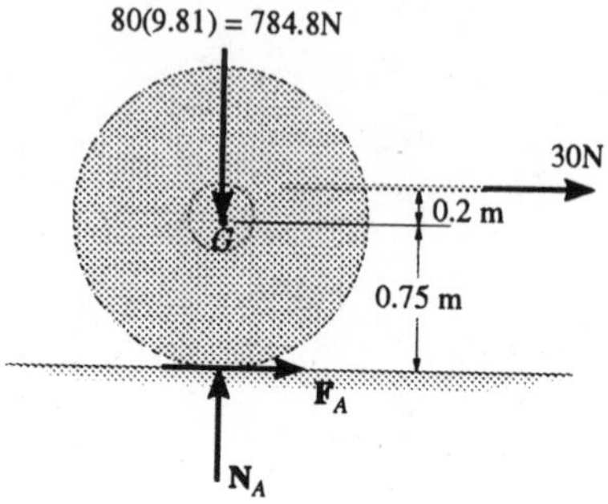

Solution

The free - body and kinetic diagrams are drawn first.

$\overset{+}{\rightarrow}\Sigma F_x = m(a_G)_x;$ ____________________

$+\uparrow \Sigma F_y = m(a_G)_y;$ ____________________

$\circlearrowleft + \Sigma M_G = I_G\alpha;$ ____________________

Assume no slipping occurs. Then

$a_G =$ ____________________

Therefore

$N_A = 785$ N

$\alpha = 0.546$ rad/s^2 **Ans.**

$F_A = 2.76$ N

Since

$(F_A)_{max} =$ ____________ $= 314$ N $>$ ____________

No slipping as assumed.

17 - 12. If the cable CB is horizontal and the beam is at rest in the position shown, determine the tension in the cable at the instant the towing force of 1500 N is applied. The coefficient of kinetic friction between the beam and the floor at A is $\mu_k = 0.3$. For the calculation, assume that the beam is a uniform

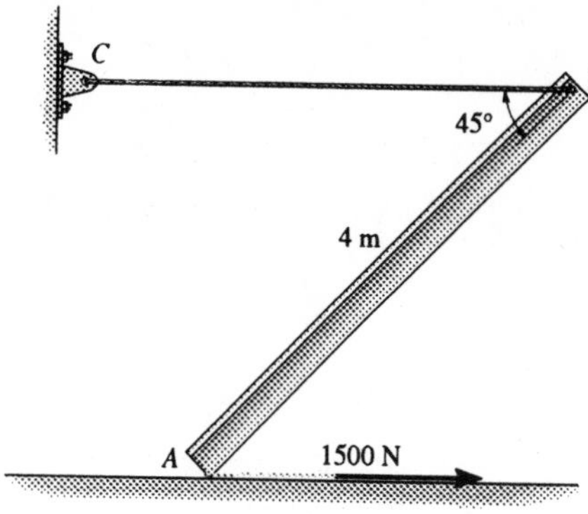

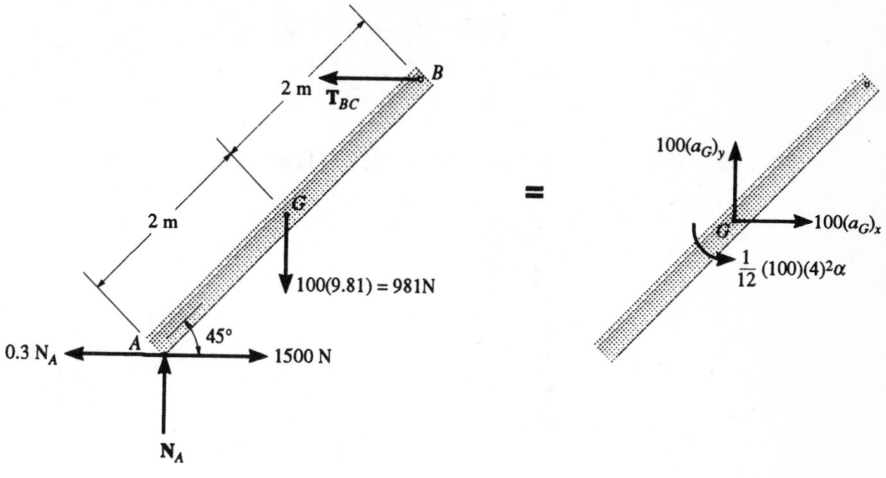

Solution

The free - body and kinetic diagrams are drawn first.

$\overset{+}{\rightarrow} \Sigma F_x = m(a_G)_x$;________________________________(1)

$\circlearrowleft + \Sigma M_G = I_G \alpha$;________________________________(2)

$+\uparrow \Sigma F_y = m(a_G)_y$;________________________________(3)

Kinematics

Apply the relative acceleration equation between points A and B, noting that $\omega = 0$. Use Cartesian vectors.

$$\mathbf{a}_B = \mathbf{a}_A + \alpha \times \mathbf{r}_{B/A} - \omega^2 \mathbf{r}_{B/A}$$

____

$$0 = a_A - \alpha\, 4 \sin 45°$$

$$a_A = 2.828\alpha$$

Using this result, apply the relative acceleration equation between points A and G.

$$\mathbf{a}_G = \mathbf{a}_A + \alpha \times \mathbf{r}_{G/A} - \omega^2 \mathbf{r}_{G/A}$$

$(a_G)_x \mathbf{i} + (a_G)_y \mathbf{j} =$________________________________

$(\overset{+}{\rightarrow})\quad (a_G)_x = 2.828\alpha - 2\alpha \sin 45° = 1.414\alpha$

$(+\uparrow)\quad (a_G)_y = 2\alpha \cos 45° = 1.414\alpha$

Substituting into Eqs. (1) - (3) yields

$T_{BC} = 636$ N **Ans.**

18 Planar Kinetics of a Rigid Body: Work and Energy

Principle of Work and Energy

18 - 1. The 50 - kg bar falls from rest. Determine the speed of its end B just before it strikes the ground.

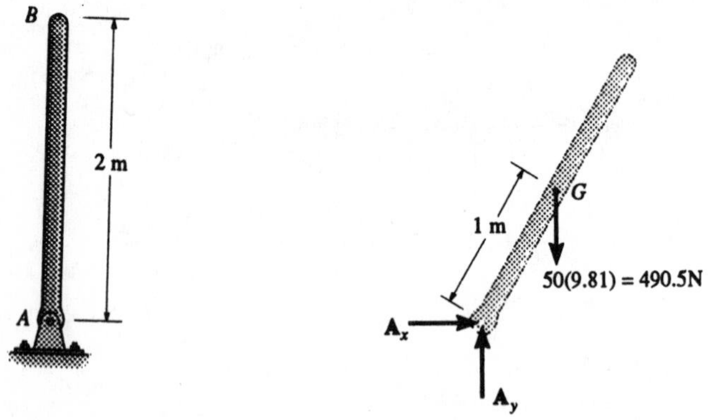

Solution

The free - body diagram is drawn first. Only the weight does work.

$$T_1 + \Sigma U_{1-2} = T_2$$

____

$$\omega = 3.84 \text{ rad/s}$$

The velocity of point B is thus,

v_B = ______________________ = 7.67 m/s **Ans.**

18 - 2. A motor supplies a constant torque or twist of $M = 200\ \text{N}\cdot\text{m}$ to the drum. If the drum has a mass of 30 kg and a radius of gyration of $k_O = 0.35$ m, determine the speed of the 15 - kg block A after it rises $s = 2\text{m}$ starting from rest. Neglect the mass of the cord.

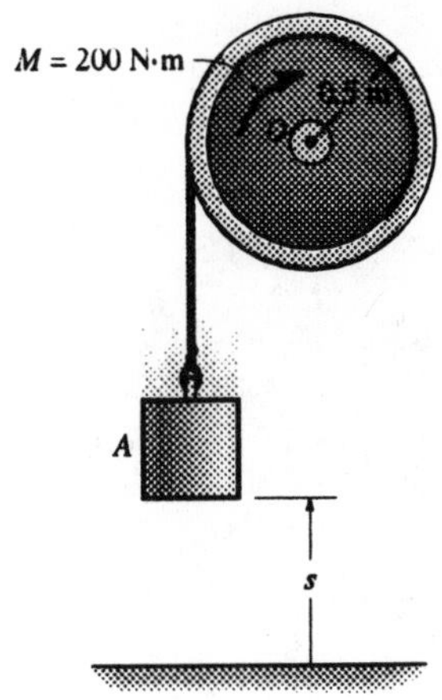

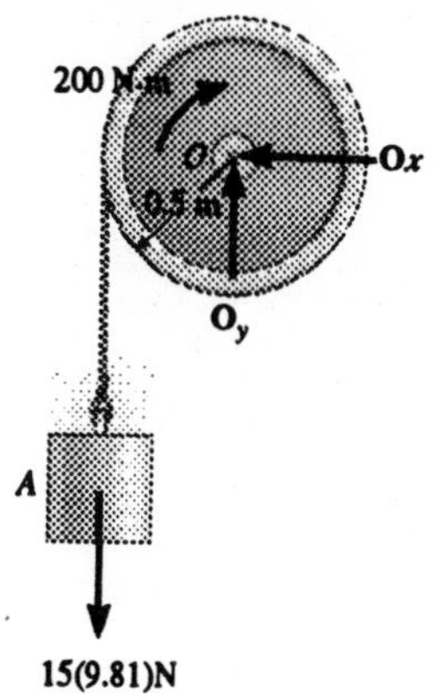

Solution

The free - body diagram of the system consisting of both the drum and block is drawn first. Only the weight of the block and the constant moment do work. When the block rises 2 m, the drum must turn

$\theta =$ ________________________ $= 4$ rad

Also, the angular velocity of the drum is $\omega = v/0.5$, where v is the velocity of the crate.

$$T_1 + \Sigma U_{1-2} = T_2$$

____

$v = 5.84$ m/s **Ans.**

18 - 3. The spool of cable, originally at rest, has a mass of 250 kg and a radius of gyration of $k_G = 350$ mm. If the spool rests on two small rollers *A* and *B* and a constant horizontal force of $P = 400$ N is applied to the end of the cable, compute the angular velocity of the spool when 2 m of cable have unraveled. Neglect friction and the mass of the rollers and unraveled cable.

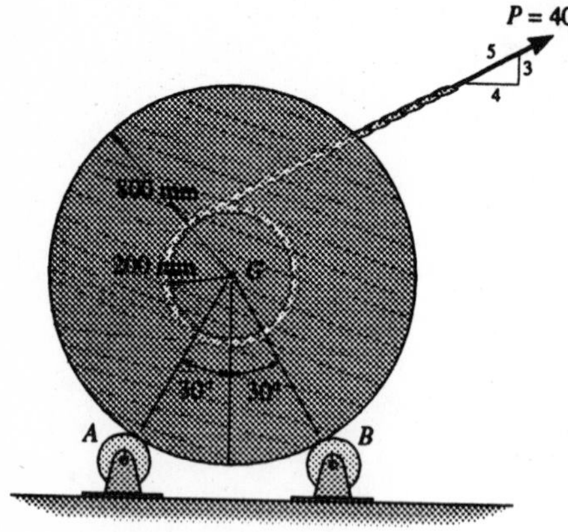

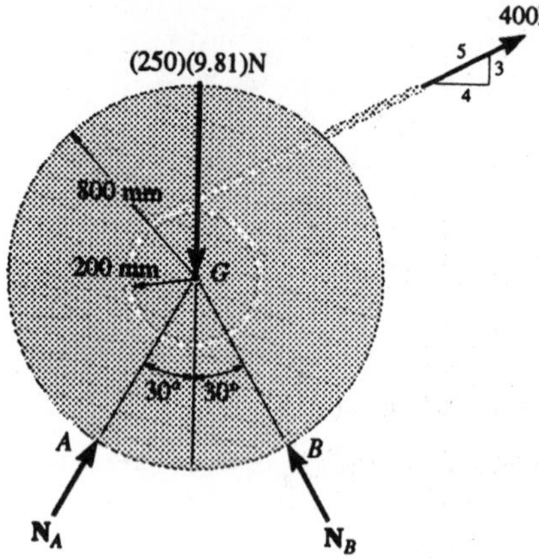

Solution

The free - body diagram is drawn first. Only the 400 N force does work.

$$T_1 + \Sigma U_{1-2} = T_2$$

$$\omega = 7.23 \text{ rad/s} \qquad \textbf{Ans.}$$

18 - 4. If the 2 - kg solid ball is released from rest when $\theta = 30°$, determine its angular velocity when $\theta = 0°$. The ball does not slip as it rolls.

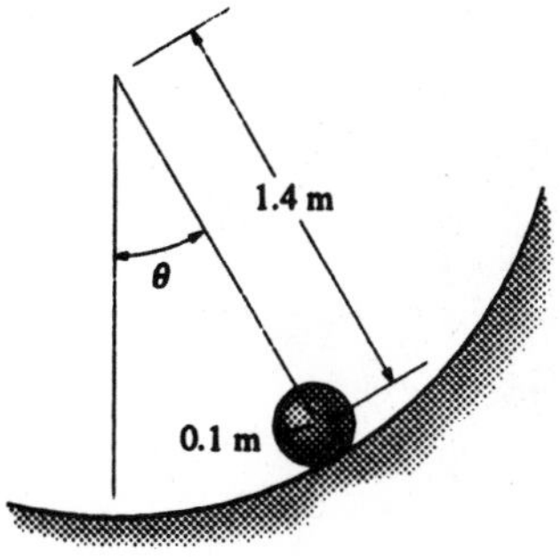

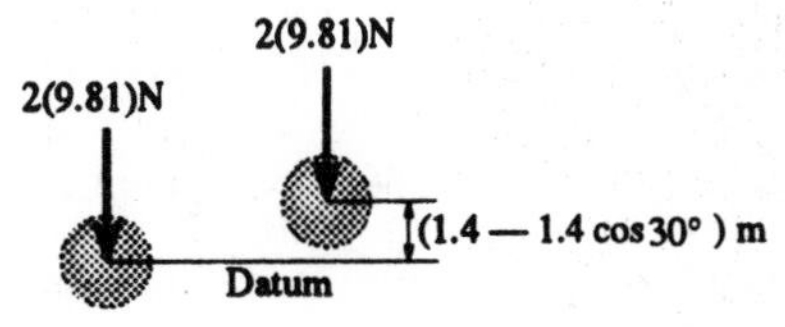

Solution

The datum is established at the lowest point. For a ball $I_G = \frac{2}{5}mr^2$. Since there is no slipping, $v_G = 0.1\omega$.

$$T_1 + V_1 = T_2 + V_2$$

$\omega = 16.2$ rad/s **Ans.**

18 - 5. The 10 - kg bar is released from rest when $\theta = 0°$. Determine its angular velocity when $\theta = 90°$. The spring has an unstretched length of 1 m.

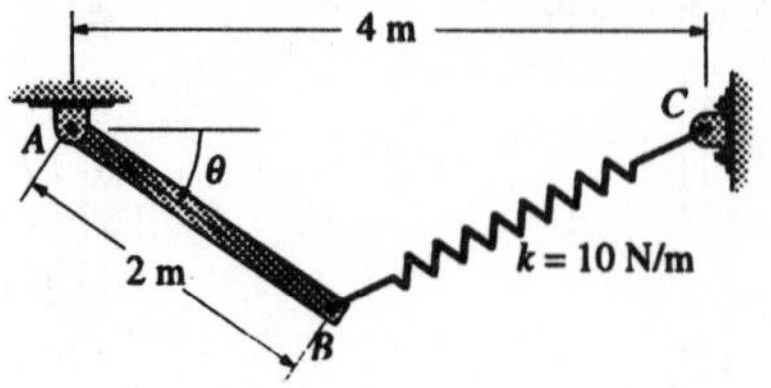

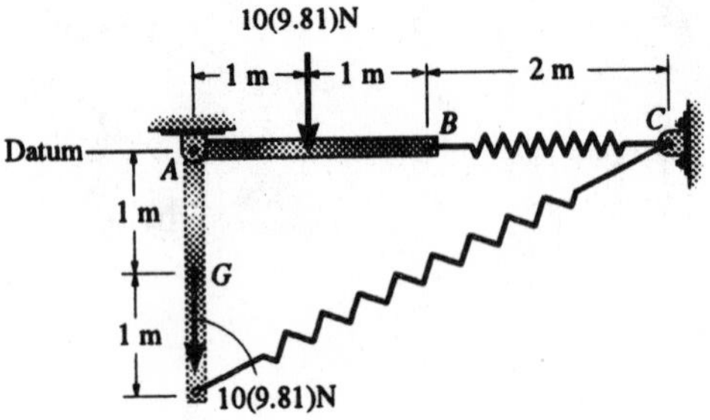

Solution

The datum is established through point A. The initial stretch of the spring is $(2-1)$ m = 1 m, and in the final position the stretch is $(\sqrt{(4)^2+(2)^2}-1)$ m = 3.472 m.

$$T_1 + V_1 = T_2 + V_2$$

$\omega = 2.53$ rad/s **Ans.**

18 - 6. The uniform slender rod has a mass of 5 kg. Determine the reaction at the pin O when the cord at A is cut and the bar swings down $\theta = 90°$.

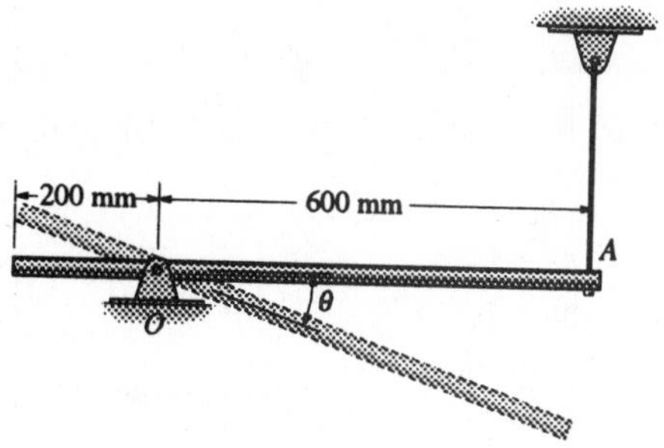

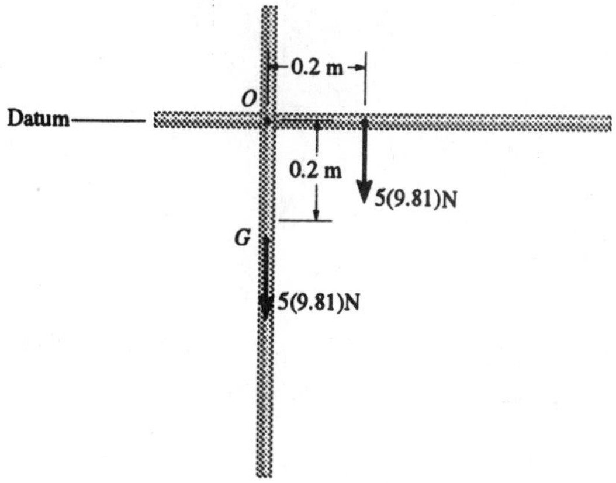

Solution

$$T_1 + V_1 = T_2 + V_2$$

$$\omega = 6.48 \text{ rad/s}$$

The free - body and kinetic diagrams are drawn first.

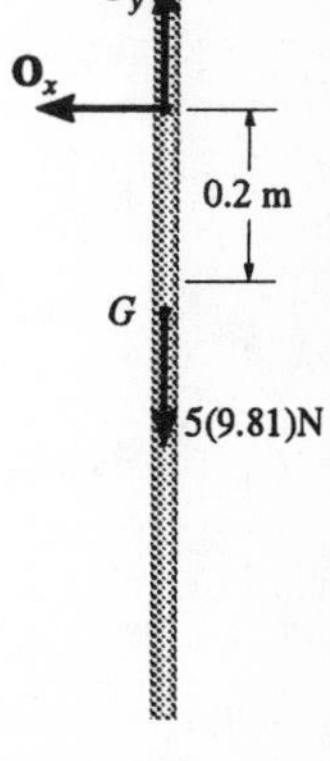

$\overset{+}{\leftarrow} \Sigma F_x = m(a_G)_x$;__________________________________

$+\uparrow \Sigma F_y = m(a_G)_y$;__________________________________

$\circlearrowleft + \Sigma M_O = I_O \alpha$;__________________________________, $\alpha = 0$

$O_x = 0 \quad O_y = 91.1 \text{ N}$ **Ans.**

19 Planar Kinetics of a Rigid Body: Impulse and Momentum

Principle of Impulse and Momentum

19 - 1. If the 10 - kg cylinder rolls without slipping at A, determine its angular velocity in 4 s starting from rest.

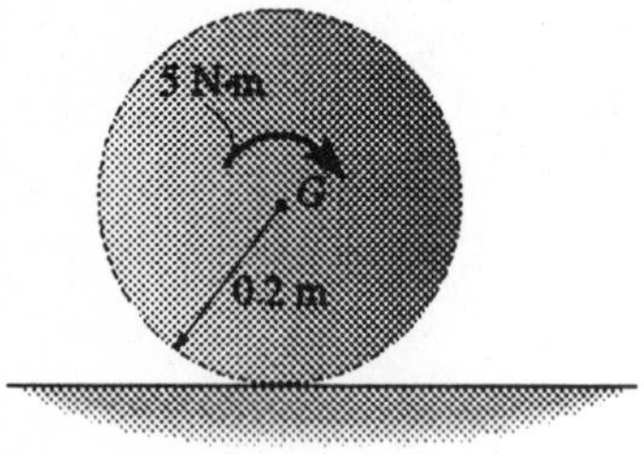

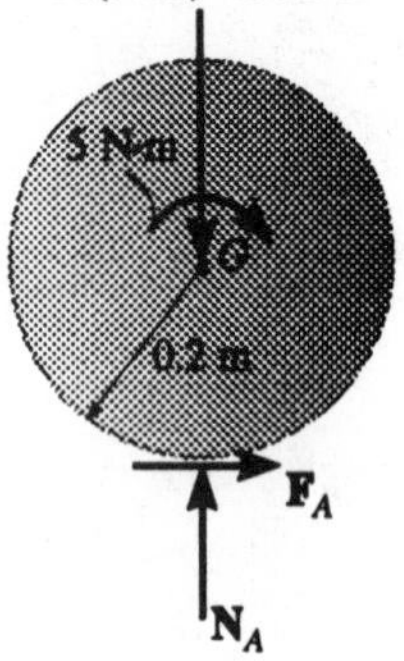

Solution

The free - body diagram of the cylinder is drawn first. Assume that after the impulses are applied the cylinder has an angular velocity ω and its mass center has a velocity of v_G.

$$(\circlearrowleft +)\ (H_G)_1 + \Sigma\int M_G dt = (H_G)_2$$

____

$$(\overset{+}{\rightarrow})\ m(v_G)_1 + \Sigma\int F_x dt = m(v_G)_2$$

____

Since no slipping occurs, relate v_G to ω.

v_G = ____________________

Solving

$\omega = 33.3$ rad/s **Ans.**

$v_G = 6.67$ m/s

19 - 2. The 50 - kg disk has an angular velocity of 30 rad/s when it is brought into contact with the horizontal surface at C. If the coefficient of kinetic friction is $\mu_k = 0.2$, determine how long it takes for the disk to stop spinning. Neglect the weight of the links.

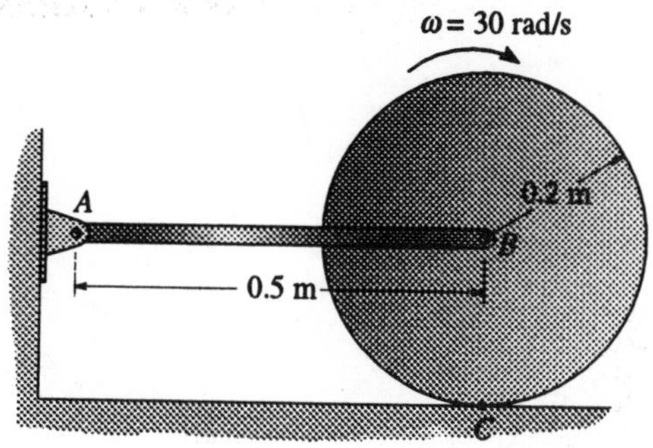

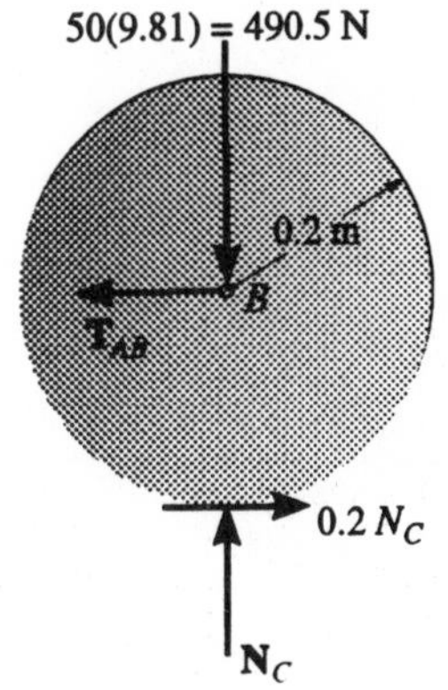

Solution

The free - diagram of the disk is drawn first.

$$(\overset{+}{\rightarrow})\ m(v_G)_{x_1} + \Sigma\int F_x\,dt = m(v_G)_{x_2}$$

$$(+\uparrow)\ m(v_G)_{y_1} + \Sigma\int F_y\,dt = m(v_G)_{y_2}$$

$$(\circlearrowleft +)\ (H_B)_1 + \Sigma\int M_B\,dt = (H_B)_2$$

Solving,

$t = 1.53$ s **Ans.**

$T_{AB} = 98.1$ N

$N_C = 490.5$ N

19 - 3. Gear A has a mass of 1.5 kg and a radius of gyration of $k_O = 0.15$ m. The coefficient of kinetic friction between the gear rack B and the horizontal surface is $\mu_k = 0.3$. If the rack has a mass of 0.8 kg and is initially moving to the left with a velocity of $(v_B)_1 = 4$ m/s, determine the constant moment **M** which must be applied to the gear to increase the motion of the rack so that in $t = 2$ s it will have a velocity of $(v_B)_2 = 8$ m/s to the left. Neglect friction between the rack and the gear and assume that the gear exerts *only* a horizontal force on the rack.

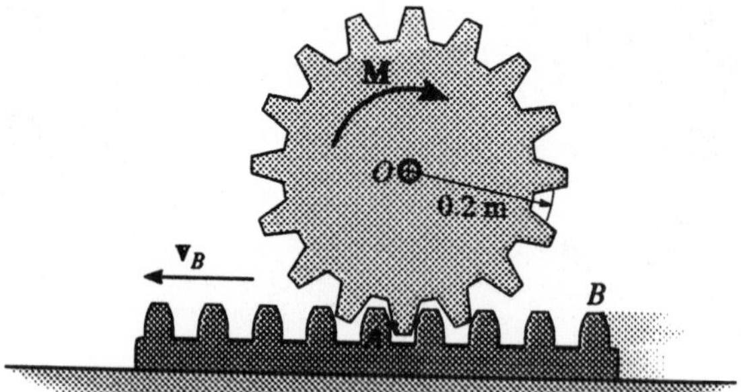

Solution

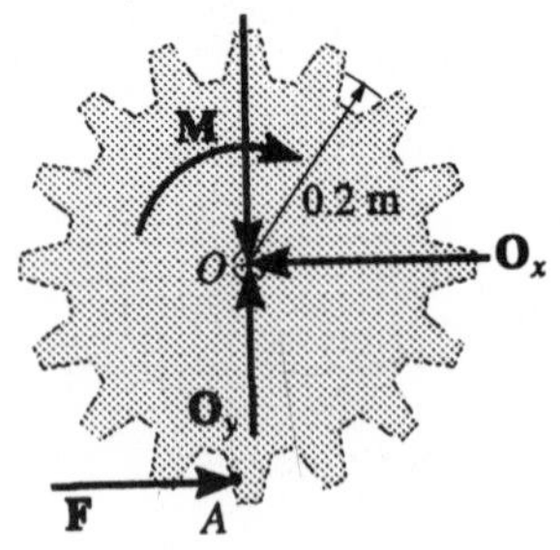

When

$(v_B)_1 = 4$ m/s, $(\omega_A)_1 =$ ____________________

$(v_B)_2 = 8$ m/s, $(\omega_A)_2 =$ ____________________

For gear A :

$$(\circlearrowleft +) \quad (H_O)_1 + \Sigma \int M_O dt = (H_O)_2$$

____

For the gear rack :

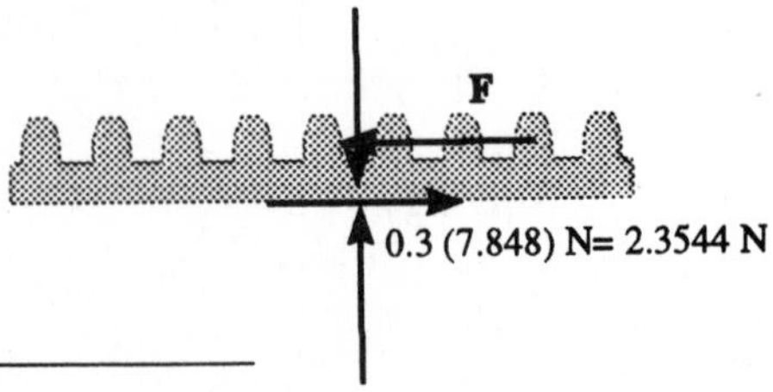

$$(\overset{+}{\leftarrow}) \quad m(v_x)_1 + \Sigma \int F_x dt = m(v_x)_2$$

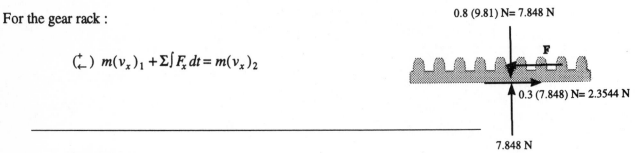

____

$F = 3.95$ N

$M = 1.13$ N · m **Ans.**

19 - 4. The 12 - kg disk has an angular velocity of $\omega = 20$ rad/s. If the brake ABC is applied such that the magnitude of force P varies with time as shown, determine the time needed to stop the disk. The coefficient of kinetic friction at B is $\mu_k = 0.4$.

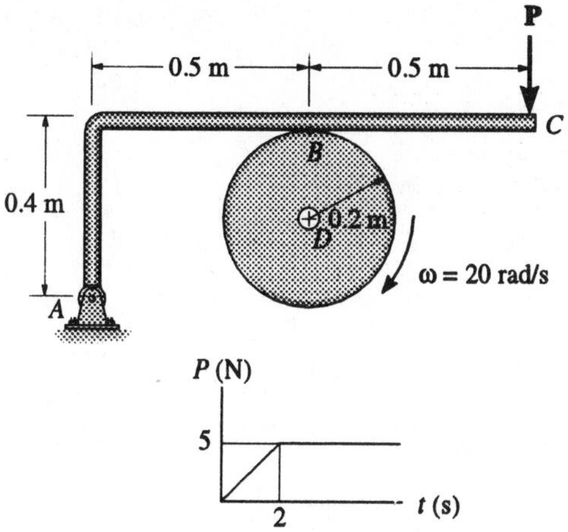

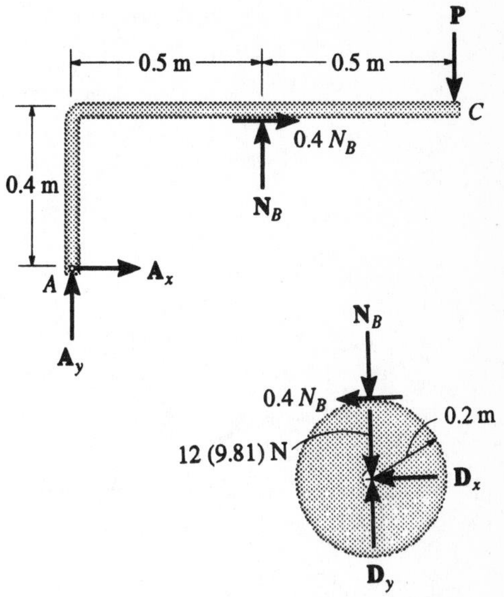

Solution

The free - body diagram of the brake and disk are drawn first.

For the brake,

$\circlearrowleft + \Sigma M_A = 0;$ ____

$$N_B = 2.941\ P$$

Using this result for the disk,

$$(\circlearrowleft +)\quad (H_D)_1 + \Sigma \int M_D dt = (H_D)_2$$

____

$$\int_0^t P dt = 20.40$$

The integral is evaluated by finding an equivalent area under the curve. Assuming $t > 2$ s,

$$\int_0^t P dt = \frac{1}{2}(5)(2) + 5(t-2) = 20.40$$

Thus,

$$t = 5.08 \text{ s}$$ **Ans.**

19 - 5. The two disks each have a mass of 30 kg and are connected together using a belt which is subjected to a tension such that it does not slip at its contacting surfaces. If a motor supplies a counter - clockwise torque, having a magnitude of $M = (12t)\ \text{N}\cdot\text{m}$, where t is in seconds, determine the angular velocity of each disk 3 s after the motor is turned on. Initially, the system is at rest.

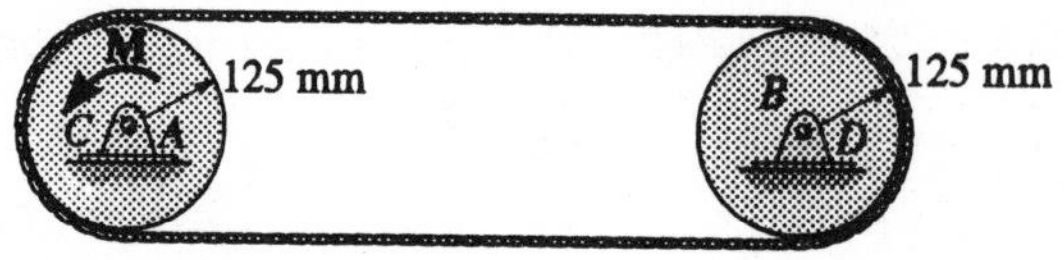

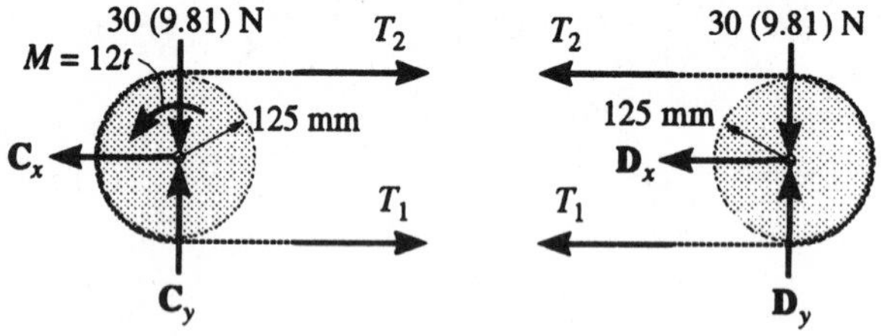

Solution

The free - body diagrams of the disks are drawn first.

For disk A :

$$(\circlearrowleft +)\quad (H_C)_1 + \Sigma\int M_C\, dt = (H_C)_2$$

For disk B :

$$(\circlearrowleft +)\quad (H_D)_1 + \Sigma\int M_D\, dt = (H_D)_2$$

Note that

$$(\omega_A)_2 = (\omega_D)_2$$

We can eliminate $(\int T_1\, dt - \int T_2\, dt)$ between the two equations and solve for the angular velocity.

$$(\omega_A)_2 = (\omega_D)_2 = 115 \text{ rad/s}$$ **Ans.**

19-6. The 10-kg bar is rotating on the smooth surface with a constant angular velocity of $\omega_1 = 3$ rad/s. Determine its new angular velocity just after its end connects to the peg P and it starts to rotate about P with out rebounding.

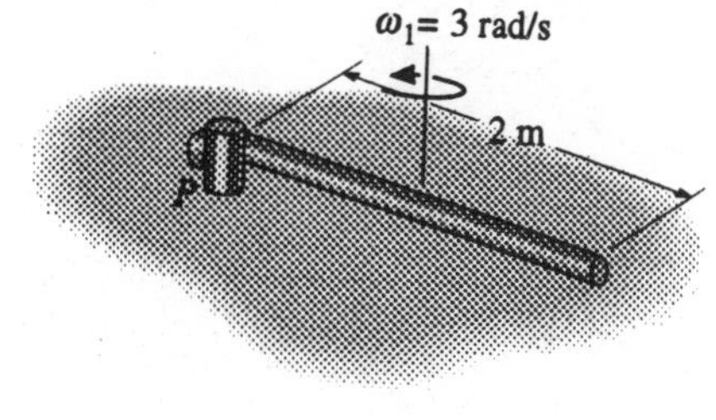

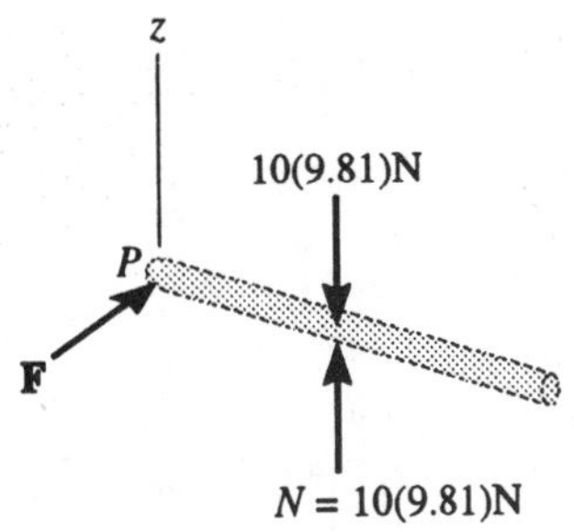

Solution

As shown on the free-body diagram, angular momentum is conserved about the z axis. Thus,

$$(H_P)_{z_1} = (H_P)_{z_2}$$

$$\omega_2 = 0.75 \text{ rad/s} \qquad \textbf{Ans.}$$

19 - 7. The 2 - kg uniform rod AB is released from rest without rotating from the position shown. As it falls, the end A strikes a hook S, which provides a permanent connection. Determine the speed at which the other end B strikes the wall.

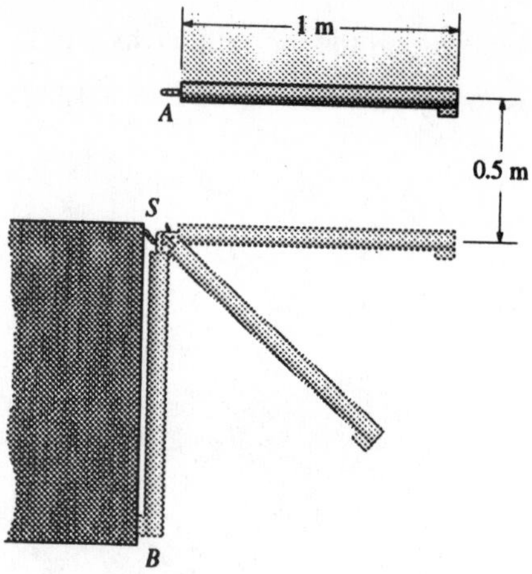

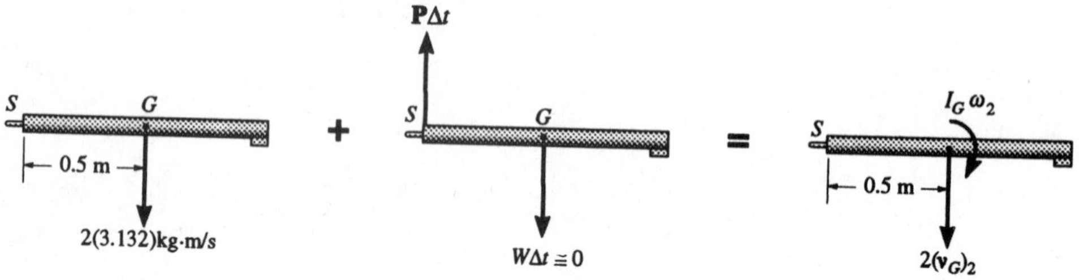

Solution

The speed of the rod's mass center just before the impact at S is determined from the conservation of energy. Putting the datum at the bar's initial position, we have

$$T_1 + V_1 = T_2 + V_2$$

____

$$v_G = 3.132 \text{ m/s}$$

Angular momentum is conserved about S by considering the weight as a nonimpulsive force. Just before and just after impact

$$(\circlearrowleft+)(H_S)_1 = (H_S)_2$$

____

$$\omega_2 = 4.698 \text{ rad/s}$$

Applying the conservation of energy theorem with the datum at the bar's horizontal position, we have

$$T_2 + V_2 = T_3 + V_3$$

____

$$\omega_3 = 7.176 \text{ rad/s}$$

Thus,

$(v_B)_3 =$ ________________________ $= 7.18$ m/s **Ans.**

19 - 8. The uniform pole has a mass of 15 kg and falls from rest when $\theta = 90°$ until it strikes the edge at A, $\theta = 60°$. If the pole then begins to pivot about this point after contact, determine the pole's angular velocity just after the impact. Assume that the pole does not slip at B as it falls until it strikes A.

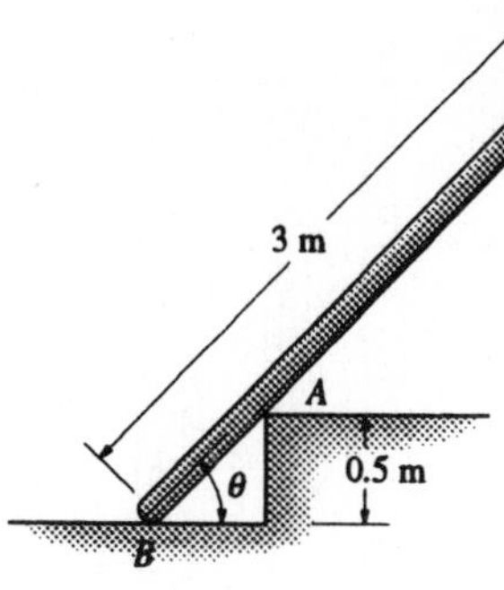

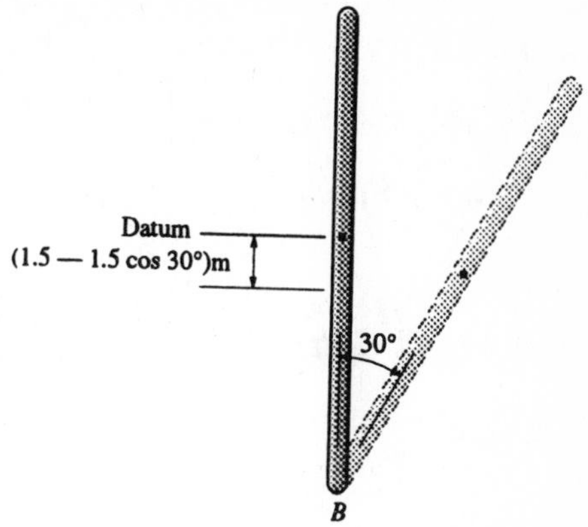

Solution

The datum is selected at B. Motion of the pole just before impact is

$$T_1 + V_1 = T_2 + V_2$$

$$\omega_2 = 1.146 \text{ rad/s}$$

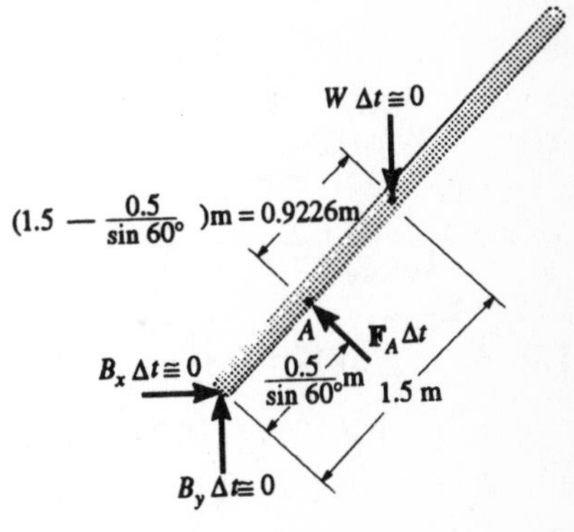

Therefore the velocity of the mass center is

$$(v_G)_2 = \underline{\hspace{6cm}} = 1.720 \text{ m/s}$$

Momentum is conserved about point A since the weight and reactive force at B can be considered nonimpulsive. Write this equation assuming the bar has a final angular velocity ω_3 and its mass center is $(v_G)_3$.

$$(\circlearrowleft +)\ (H_A)_2 = (H_A)_3$$

Since

$$(v_G)_3 = 0.9226\omega_3$$

Then

$$\omega_3 = 1.53 \text{ rad/s}$$ **Ans.**

Answers

12 - 1. $0 = 8 + (-1)t$

$0 = (8)^2 + 2(-1)(s - 0)$

$s = 0 + 8(8) + \frac{1}{2}(-1)(8)^2$

12 - 2. $\int_0^s ds = \int_0^t 2t dt$

2 m/s^2

12 - 3. $(0)^2 = (12)^2 + 2(-9.81)(h - 0)$

$0 = 12 + (-9.81)t_{AB}$

$-20 = 0 + 12t_{AC} + \frac{1}{2}(-9.81)t_{AC}{}^2$

12 - 4. $15(2)^3 - 3(2)$

$15(4)^3 - 3(4)$

$948 - 114$

$45t^2 - 3$

$90\,t$

12 - 5. $(20)^2 = 0 + 2a_1(50)$

$(30)^2 = (20)^2 + 2a_2(150 - 50)$

$20 = 0 + 4t_1$

$30 = 20 + (1.25)t^2$

$\frac{(150 - 0)}{(5 + 4)}$

$\frac{(30 - 0)}{(5 + 4)}$

12 - 6. $(10)(30) + \frac{1}{2}(10)(60 - 30)$

12 - 7. $\int_0^v dv = \int_0^t 0.8\, t dt$

$0.4(10)^2$

$\int_{40}^v dv = \int_{10}^t 8 dt$

12 - 8. $3(10)$

$3(10)+(5)(20-10)$

$\frac{1}{2}(1)(30)$

$\frac{1}{2}(10)(30)+\frac{1}{2}(30+80)(20-10)$

12 - 9. $\frac{6-0}{10-0}$

$(6-6)/(20-10)$

$\frac{1}{2}(10)(6)$

$\frac{1}{2}(10)(6)+6(20-10)$

$\int_0^s ds=\int_0^t 0.6t dt$

$\int_{30}^s ds=\int_{10}^t 6dt$

12 - 10. $0.2s(0.2ds)=a\,ds$

$(40-0.2s)(-0.2ds)=a\,ds$

12 - 11. $0.125(4)^2$

$0.03(4)^3$

$\sqrt{(2)^2+(1.92)^2}$

$\dot{x}=0.25(4)$

$\dot{y}=0.09(4)^2$

$\ddot{x}=0.25$

$\ddot{y}=0.18(t)=0.18(4)$

12 - 12. 4

0

$-4x\dot{x}$

$-4\dot{x}^2-4x\ddot{x}$

12 - 13. $v_A\cos 30°$

$v_A\sin 30°$

$25=0+v_A\cos 30° t$

$0=0+v_A\sin 30° t+\frac{1}{2}(-9.81)t^2$

12 - 14. $6 = 0 + v_A \cos\theta(2)$

$0 = (v_A \sin\theta)^2 + 2(-9.81)(h-1)$

$0 = v_A \sin\theta + (-9.81)(2)$

12 - 15. $15(\frac{3}{5})$

$15(\frac{4}{5})$

$R(\frac{4}{5}) = 0 + 9t$

$-R(\frac{3}{5}) = 0 + 12t + \frac{1}{2}(-9.81)t^2$

12 - 16. $(7)^2/60$

$-3\ \text{m/s}^2$

$\sqrt{(0.816)^2 + (-3)^2} =$

12 - 17. $(\frac{10-8}{5})$

$\frac{(9)^2}{200}$

12 - 18. $\frac{1}{2}x\Big|_{x=2} = 1$

$\frac{1}{2}$

$4\ \text{m/s}^2$

$8^2/5.66$

12 - 19. $v = 0.1t^2$

$s = 0.0333t^3$

$100(\pi/6)$

$(52.36/0.0333)^{1/3}$

$0.1(11.62)^2$

$0.2(11.62)$

$(13.51)^2/100$

$\sqrt{(2.325)^2 + (1.826)^2}$

12 - 20. 9.81 m/s^2

0

$(50)^2/9.81$

50 m/s

$(v_B)_y^2 = 0 + 2(9.81)(500 - 0)$

$\sqrt{(50)^2 + (99.05)^2}$

$9.81\cos 63.21°$

$9.81\sin 63.21°$

$(110.95)^2/4.42$

12 - 21. $e^{\theta}\dot{\theta}$

$e^{\theta}\dot{\theta}^2 + e^{\theta}\ddot{\theta}$

$4\theta\dot{\theta} + 8\cos\theta\,\dot{\theta}$

$4\dot{\theta}^2 + 4\theta\ddot{\theta} - 8\sin\theta\,\dot{\theta}^2 + 8\cos\theta\,\ddot{\theta}$

$8\cos 2\theta\,\dot{\theta}$

$-16\sin 2\theta\,\dot{\theta}^2 + 8\cos 2\theta\,\ddot{\theta}$

12 - 22. 0

0

$\sqrt{(0)^2 + (40)^2}$

$\sqrt{(-16)^2 + (20)^2}$

12 - 23. 25 m

0

0

$0.3t$

0.3 rad/s

0

$-8\cos(0.3t)$

$8(0.3)\sin(0.3t)$

$8(0.3)^2\cos(0.3t)$

12 - 24. 3

0

$500\sec\theta\tan\theta\dot{\theta}$

$500[(\sec\theta\tan^2\theta + \sec^3\theta)\dot{\theta}^2 + \sec\theta\tan^2\theta\,\ddot{\theta}]$

12 - 25. 0.25

0

0

4 rad/s

0

$0.25\cos\theta$

$-0.25\sin\theta\,\dot{\theta}$

$-0.25[\sin\theta\,\ddot{\theta}+\cos\theta\,\dot{\theta}^2]$

12 - 26. $3s_B+s_A=l$

$3v_B=-v_A$

12 - 27. $2s_C+s_B=l$

$s_A+(h-s_C)=l'$

$2v_C=-v_B$

$v_A=v_C$

12 - 28. $4(x_B)=4(5)=20$ m

$3x_B+\sqrt{x_D^2+(5)^2}$

$3\dot{x}_B+\frac{1}{2}[x_D^2+(5)^2]^{-1/2}(2x_D\dot{x}_D)=0$

$0+2(2)(3-0)$

12 - 29. $2s_H+s_P=l$

$2v_H=-v_P$

$12=-6+v_{P/H}$

12 - 30. $-20\cos45°\mathbf{i}+20\sin45°\mathbf{j}=45\mathbf{i}+(v_{A/B})_x\mathbf{i}+(v_{A/B})_y\mathbf{j}$

$45+(v_{A/B})_x$

$0+(v_{A/B})_y$

$\frac{(20)^2}{50}\cos45°\mathbf{i}+\frac{(20)^2}{50}\sin45°\mathbf{j}=4\mathbf{i}+(a_{A/B})_x\mathbf{i}+(a_{A/B})_y\mathbf{j}$

12 - 31. $-18\mathbf{j}=25\mathbf{i}+(v_{A/B})_x\mathbf{i}+(v_{A/B})_y\mathbf{j}$

$\frac{(18)^2}{300}\mathbf{i}+1.5\mathbf{j}=2\mathbf{i}+(a_{A/B})_x\mathbf{i}+(a_{A/B})_y\mathbf{j}$

13-1. $7500 - 5500 = 72\,000a$

13-2. $2(9.81) - 20 = 2a$

13-3. $0.8t$

$T = 300(4)$

$\int_0^s ds = \int_0^5 0.4t^2\,dt$

13-4. $N_B - 2(9.81)\cos 30° = 0$

$-0.3(16.99) + 2(9.81)\sin 30° = 2a$

$v_1^2 = 0 + 2(2.356)(3-0)$

13-5. $3000 - 1500s = 4a$

$\int_0^{0.05} (750 - 37.5s)\,ds = \int_0^v v\,dv$

13-6. $196.2 - 2T = 20a_A$

$-T = 10a_B$

$2s_A + s_B = l$

13-7. $19.62 = 2(\frac{v_A{}^2}{2})$

No, water will fall out tangent to the path of motion.

13-8. $N_T - 100(9.81)\cos 72.6° = 100(\frac{(4)^2}{94.2})$

$100(9.81)\sin 72.6° = 100a_t$

13-9. $2(9.81)\sin\theta = 2a_t$

$N_B - 2(9.81)\cos\theta = -2(\frac{v_A{}^2}{.2})$

$\int_0^{v_A} v\,dv = 9.81\int_{0°}^{30°} \sin\theta\,(2d\theta)$

13-10. $N_C \cos 30° - 2(9.81)\sin 30°$

$N_C \sin 30° + F_C - 2(9.81)\cos 30°$

0

$-1.4\sin\theta\dot{\theta}$

$-1.4(\sin\theta\,\ddot{\theta} + \cos\theta\,\dot{\theta}^2)$

13 - 11. $2\theta/2$

$-N_C \sin 57.52°$

$P+N_C \cos 57.52°$

3 rad/s

0

$2\dot{\theta} = 2(3) = 6$

$2\ddot{\theta} = 0$

13 - 12. $0.15 \sin\theta$

$$\frac{0.15(2-\cos 90°)}{0.15 \sin 90°}$$

$-N_C \cos 26.57°$

$N_C \sin 26.57° + F$

14 - 1. $\frac{1}{2}(2000)(8.33)^2 - F_{avg}(0.2) = 0$

14 - 2. $\frac{1}{2}800(19.44)^2 - [40(10^3)(0.75) + 60(10^3)(s-0.75)] = 0$

14 - 3. $\frac{1}{2}(2)(2)^2 + 19.62(3-1) = \frac{1}{2}(2)v_B^2$

$-1 = 0 + 3.288t + \frac{1}{2}(-9.81)t^2$

$5.694(0.897)$

14 - 4. $\frac{1}{2}(2)(2)^2 + 40(0.75+s) - 3.924(0.75+s)$

$-\frac{1}{2}(800)s^2 - \frac{1}{2}(400)(s-0.1)^2 = 0$

14 - 5. $490.5(6)$

$2943/3$

$981/4(10^3)$

14 - 6. $19.62 \sin 15°$

$5.078(20)$

14 - 7. $3(400) - 100(9.81) = 100a$

$v = 0 + 2.19(3)$

$3s_B - s_P = l$

$v_P = 3v_B = 3(6.57)$

$400(19.71)$

14 - 8. $0+[\frac{1}{2}(40)(4)^2] = \frac{1}{2}(2)(v_2)^2 + \frac{1}{2}(40)(3)^2$

14 - 9. $2s_A + s_B = l$

$0+0+0+0 = \frac{1}{2}(50)(v_A)^2 + \frac{1}{2}(10)(-2v_A)^2 - 50(9.81)(2) + 0$

14 - 10. $0+0+0 = 0 - 5(9.81)(0.4+s_A) + \frac{1}{2}(800)(s_A)^2 + \frac{1}{2}(400)(s_A - 0.05)^2$

15 - 1. $-(0.2)(10) + \int F_x\,dt = (0.2)(20)\cos 40°$

$0 + \int F_y\,dt = (0.2)(20)\sin 40°$

$\sqrt{(5.06)^2 + (2.57)^2}$

15 - 2. $10(3) + \int_0^2 30\cos(\frac{\pi}{4}t)\,dt = 10\,v_2$

15 - 3. $\frac{147.2}{60}$

$0 + \int_{2.45}^{5} 60t\,dt + 300(6-5) - 147.2(6-2.45) = 15\,v_2$

15 - 4. $0 - T_A(2) + 9.81(2) = 10(v_A)_2$

$0 - 49.05(2) + T_B(2) = 5(v_B)_2$

$T_A = T_B$

$2s_A + s_C = l$

$(h - s_C) + (h - s_B) = l'$

15 - 5. $0+0 = (0.0015)(1400) - 2.5(v_R)_2$

15 - 6. $0.6(10)\cos 30° = (25.6)v$

15 - 7. $0+0 = -80v_A + 70(2 - v_A)$

$70(2 - 0.933) = 150v_B$

15 - 8. $0+0 = 8[2\cos 30° - v_m] - 70(v_m)_2$

$0+0+N_{avg}(1.5) - (70+8)(9.81)(1.5) = 8(2\sin 30°)$

15 - 9. $(0.5)(2) + 0 = (0.5)(v_B)_2 + 2(v_A)_2$

$\frac{(v_A)_2 - (v_B)_2}{2-0} = 0.4$

$\frac{1}{2}(2)(0.560)^2 - 7.848x = 0$

15-10. $0=[v_C \sin 20°]^2+2(-9.81)(4-$ ...

$22.43 \cos 20°$

$3.5(10^3)(21.08)+0=7(10^3)(v_B)_2$

$4=0+0+\frac{1}{2}(9.81)t^2$

$0+(10.54)(0.903)$

15-11. $4(4)+0=4(v_A)_2+4(v_B)_2$

$0.7=\frac{(v_B)_2-(v_A)_2}{4-0}$

$\frac{1}{2}(4)(3.40)^2+0=0+\frac{1}{2}(500)x^2$

15-12. $-2(4\cos 30°)+2(2\sin 30°)=2(v_A)_{2x}+2(v_B)_{2x}$

$0.65=\frac{(v_B)_{2x}-(v_A)_{2x}}{-4\cos 30°-2\sin 30°}$

$2(-4\sin 30°)=2(v_A)_{2y}$

$2(-2\cos 30°)=2(v_B)_{2y}$

15-13. $2[0.5(2)(0.4)]+\int_0^2(0.5t)\,dt=2[0.5(v_2)(0.4)]$

15-14. $\frac{1}{2}(50)(2.5)^2+0=\frac{1}{2}(50)(v_B)^2-50(9.81)(1)$

$50(2.5)(3)=50(v_B)_{\text{horiz.}}(2)$

$\sqrt{(5.09)^2-(3.75)^2}$

16-1. $\int_5^{\omega}\omega\,d\omega=\int_0^{4\pi}0.2\theta\,d\theta$

$7.52(0.5)$

16-2. $0.3(4)\cos 4t$

1.2 rad/s

$-4(1.2)\sin 4t$

0

$1.2(250)$

0

$(1.2)^2(200)$

16 - 3. $\int_0^{\omega} d\omega = \int_0^{t} 3t^2 dt$

$\int_0^{\theta} d\theta = \int_0^{t} t^3 dt$

$3(2)^2$

$(2)^3$

$\frac{1}{4}(2)^4$

$8(0.20)$

$(8)^2(0.2)$

$12(0.2)$

$\sqrt{(12.8)^2 + (2.4)^2}$

16 - 4. $0.2 \cos\theta$

$-0.2 \sin\theta\, \dot{\theta}$

$-0.2 \cos\theta\, (\dot{\theta})^2 - 0.2 \sin\theta\, \ddot{\theta}$

16 - 5. $0.3 \cot\theta$

$-0.3 \csc^2\theta\, \dot{\theta}$

16 - 6. $2r\theta$

$2r\omega$

16 - 7. $2(1.5 \sin\theta)$

$3 \cos\theta\, \dot{\theta}$

$3 \cos\theta$

$3 \sin\theta\, \dot{\theta}$

16 - 8. $3 \cos\theta$

$-3 \sin\theta\, \dot{\theta}$

$3 \sin\theta$

$3 \cos\theta\, \dot{\theta}$

16 - 9. $200\mathbf{j} + (3\mathbf{k}) \times (12\mathbf{j})$

$\sqrt{(36)^2 + (200)^2}$

16 - 10. $\omega_{AB}(1)$

$-\omega_{AB}(1)\mathbf{j} = -4\mathbf{j} + (-\omega\mathbf{k}) \times (-1.5 \cos 30°\mathbf{i} - 1.5 \sin 30°\mathbf{j})$

16 - 11. $4(0.4)$

$-(\frac{4}{5})v_B\mathbf{i}-(\frac{3}{5})v_B\mathbf{j}=-1.6\mathbf{i}+(\omega_{BC}\mathbf{k})\times(-1\mathbf{i})$

$\frac{2}{0.5}$

16 - 12. $v_A\mathbf{i}=-6\mathbf{j}+(\omega\ \mathbf{k})\times(0.3\mathbf{i}-0.4\mathbf{j})$

16 - 13. $8(75)$

$-v_C\mathbf{i}=-600\mathbf{i}+(\omega_{BC}\mathbf{k})\times(100\cos30°\mathbf{i}+100\sin30°\mathbf{j})$

$\frac{600}{150}=4\text{ rad/s}$

$\frac{400}{25}$

$4(100)=400\text{ mm/s}$

16 - 14. $\frac{12}{0.350}$

$(34.29)(0.7)$

16 - 15. $4(0.15)$

$\frac{0.6}{0.1}$

$6(0.05)$

$\frac{3}{0.15+0.05}$

16 - 16. 0

6

6/2

16 - 17. $\frac{3}{0.075}$

$(a_C)_x\mathbf{i}-4\mathbf{j}=-(a_D)_y\mathbf{j}+(\alpha\mathbf{k})\times(-0.075\mathbf{i})-(40)^2(-0.075\mathbf{i})$

16 - 18. $(a_A)_x\mathbf{i}-(a_A)_y\mathbf{j}=0.6\mathbf{i}+(-4\mathbf{k})\times(0.15\mathbf{j})-(2)^2(0.15\mathbf{j})$

$(a_B)\mathbf{i}=1.20\mathbf{i}-0.6\mathbf{j}+(\alpha_{AB}\mathbf{k})\times(0.4\mathbf{i}-0.3\mathbf{j})-\mathbf{0}$

16 - 19. $0.5(0.9)$

0.45

$\overline{0.2/\sin 30^\circ}$

$1.125(\frac{0.2}{\tan 30^\circ})$

$0.390/1.25$

$2(0.9)$

$(0.5)^2(0.9)$

$(0.312)^2(1.25)$

$-(a_C)_x\mathbf{i} - 0.122\mathbf{j} = -1.80\cos 30^\circ\mathbf{i} + 1.80\sin 30^\circ\mathbf{j}$

$-0.225\sin 30^\circ\mathbf{i} - 0.225\cos 30^\circ\mathbf{j}$

$+(\alpha_{BC}\mathbf{k})\times(0.2\mathbf{i}) - (1.125)^2(0.2\mathbf{i})$

$1.92/1.25$

17 - 1. $N_y - 98.1 = -10(4\sin 30^\circ)$

$V_x = 10(4\cos 30^\circ)$

$M_A = 10(4\cos 30^\circ)(1)$

17 - 2. $1.8v^2 = 1300a_G$

$(0.75)(1.8v^2) - 1.25N_A = 0$

$N_A - (1300)(9.81) = 0$

17 - 3. $N - 98.1 = -10(6)$

$0.4(38.1) = 10\alpha(1.5)$

17 - 4. 0

$\alpha(1)$

$0.4N_C = 100(1\alpha)\cos 30^\circ$

$N_C - 981 = 100(1\alpha)\sin 30^\circ$

17 - 5. $A_x = 4000(3) + 800(3)$

$-14\,400(1.75) - A_y(8) + 4000(9.81)(4) + 800(9.81)(2.5) = -4000(3)(1.5) - 800(3)(1.75)$

17 - 6. $-147.15(1.5) = [\frac{1}{3}(15)(3)^2]\alpha$

$A_x = 15(5)^2(1.5)$

$A_y - 147.15 = -15(4.905)(1.5)$

17-7. $A_x = 30(0.2\sqrt{2}\alpha)(\sin 45°)$

$A_y - 294.3 = -30(0.2\sqrt{2}\alpha)(\cos 45°)$

$294.3(0.2) = 0.8\alpha + [30(0.2\sqrt{2}\alpha)]0.2\sqrt{2}$

17-8. $\mu_A N_A = 80(2.5\alpha)\cos\theta - 80(2.5\omega^2)\sin\theta$

$N_A - 784.8 = -80(2.5\omega^2)\cos\theta - 80(2.5\alpha)\sin\theta$

$-784.8\sin\theta\,(2.5) = -[80(1.2)^2]\alpha - 80(2.5\alpha)(2.5)$

$\int_0^\theta 3.19\sin\theta d\theta = \int_0^\omega \omega\, d\omega$

17-9. $F_A = 90(0.52\alpha)(\frac{12}{13})$

$N_A - 882.9 = -90(0.52\alpha)(\frac{5}{13})$

$882.9(0.2) = [90(0.45)^2]\alpha + 90(0.52\alpha)(0.52)$

17-10. $30 = 200a_G$

$T - 1962 = 0$

$30(1.25) = [\frac{1}{12}(200)(4)^2]\alpha$

17-11. $30 + F_A = 80a_G$

$N_A - 784.8 = 0$

$F_A(0.75) - 30(0.2) = -[80(0.3)^2]\alpha$

0.75α

$0.4(785\text{ N}) > 2.76\text{ N}$

17-12. $1500 - T_{BC} - 0.3N_A = 100(a_G)_x$

$1500(2\sin 45°) - 0.3N_A(2\sin 45°) - N_A(2\cos 45°) + T_{BC}(2\sin 45°) = [\frac{1}{12}(100)(4)^2]\alpha$

$N_A - 981 = 100(a_G)_y$

$a_B\mathbf{j} = a_A\mathbf{i} + \alpha\mathbf{k} \times (4\cos 45°\mathbf{i} + 4\sin 45°\mathbf{j})$

$2.828\alpha\mathbf{i} + \alpha\mathbf{k} \times (2\cos 45°\mathbf{i} + 2\sin 45°\mathbf{j})$

18-1. $0 + 490.5(1) = \frac{1}{2}[\frac{1}{3}(50)(2)^2]\omega^2$

18-2. 210.5

$(0+0) + 200(4) - 15(9.81)(2) = [\frac{1}{2}(15)(v)^2] + [\frac{1}{2}(30)(0.35)^2](\frac{v}{0.5})^2$

18 - 3. $0+(400)(2)=\frac{1}{2}[250(0.350)^2](\omega_2)^2$

18 - 4. $0+2(9.81)[(1.4)-(1.4)\cos 30°]=\frac{1}{2}(2)(0.1\omega)^2+\frac{1}{2}[\frac{2}{5}(2)(0.1)^2]\omega^2$

18 - 5. $0+0+\frac{1}{2}(10)(1)^2=\frac{1}{2}[\frac{1}{3}(10)(2)^2]\omega^2-10(9.81)(1)+\frac{1}{2}(10)(3.472)^2$

18 - 6. $0+0=\frac{1}{2}[\frac{1}{12}(5)(0.8)^2]\omega^2+\frac{1}{2}(5)(0.2\omega)^2-5(9.81)(0.2)$

$O_x=5(0.2\alpha)$

$O_y-5(9.81)=5(6.48)^2(0.2)$

$0=-[\frac{1}{12}(5)(0.8)^2+5(0.2)^2]\alpha$

19 - 1. $0+5(4)-F_A(0.2)(4)=[\frac{1}{2}(10)(0.2)^2]\omega$

$0+F_A(4)=10v_G$

$v_G=0.2\omega$

19 - 2. $0+0.2N_C(t)-T_{AB}(t)=0$

$0+N_C(t)-490.5(t)=0$

$[\frac{1}{2}(50)(0.2)^2]30-0.2N_C(t)(0.2)=0$

19 - 3. $4/0.2=20 \text{ rad/s}$

$8/0.2=40 \text{ rad/s}$

$[1.5(0.15)^2](20)+M(2)-F(2)(0.2)=[1.5(0.15)^2](40)$

$(0.8)(4)+F(2)-2.354(2)=0.8(8)$

19 - 4. $N_B(0.5)-P(1)-0.4N_B(0.4)=0$

$[\frac{1}{2}(12)(0.2)^2]20-\int_0^t 0.4(2.941P)(0.2)dt=0$

19 - 5. $0+\int_0^3 12t\,dt+0.125\int T_1\,dt-0.125\int T_2\,dt=$

$[\frac{1}{2}(30)(0.125)^2](\omega_A)_2$

$0+(0.125)\int T_2\,dt-(0.125)\int T_1\,dt=$

$[\frac{1}{2}(30)(0.125)^2](\omega_D)_2$

19 - 6. $[\frac{1}{12}(10)(2)^2]3 = [\frac{1}{3}(10)(2)^2]\omega_2$

19 - 7. $0+0 = \frac{1}{2}(2)(v_G)^2 - 2(9.81)(0.5)$

$(2)(3.132)(0.5) = [\frac{1}{3}(2)(1)^2]\omega_2$

$\frac{1}{2}[\frac{1}{3}(2)(1)^2](4.698)^2 + 0 = \frac{1}{2}[\frac{1}{3}(2)(1)^2](\omega_3)^2 - 2(9.81)(0.5)$

7.176(1)

19 - 8. $0+0 = \frac{1}{2}[\frac{1}{3}(15)(3)^2]\ \omega_2^2 - (15)(9.81)[1.5 - 1.5\cos 30°]$

$[\frac{1}{12}(15)(3)^2](1.146) + 15(1.720)(0.9226)$

$= [\frac{1}{12}(15)(3)^2]\ \omega_3 + 15(v_G)_3(0.9226)$